Place, Diversity and Solidarity

In many countries, particularly in the Global North, established forms of solidarity within communities are said to be challenged by the increasing ethnic and cultural diversity of the population. Against the backdrop of renewed geopolitical tensions – which inflate and exploit ethno-cultural, rather than political-economic cleavages – concerns are raised that ethnic and cultural diversity challenge both the formal mechanisms of redistribution and informal acts of charity, reciprocity and support which underpin common notions of community.

This book focuses on the innovative forms of solidarity that develop around the joint appropriation and the envisaged common future of specific places. Drawing on examples from schools, streets, community centres, workplaces, churches, housing projects and sporting projects, it provides an alternative research agenda from the 'loss of community' narrative. It reflects on the different spatiotemporal frames in which solidarities are nurtured, the connections forged between solidarity and citizenship, and the role of interventions by professionals to nurture solidarity in diversity.

This timely and original work will be essential reading for those working in human geography, sociology, ethnic studies, social work, urban studies, political studies and cultural studies.

Stijn Oosterlynck is Associate Professor in Urban Sociology at the University of Antwerp, Belgium.

Nick Schuermans is a postdoctoral researcher at the Centre on Inequalities, Poverty, Social Exclusion and the City of the University of Antwerp and a teaching associate at the geography department of the Vrije Universiteit Brussel, Belgium.

Maarten Loopmans is Associate Professor at the Division of Geography at KU Leuven, Belgium.

Routledge Studies in Human Geography

This series provides a forum for innovative, vibrant and critical debate within Human Geography. Titles will reflect the wealth of research which is taking place in this diverse and ever-expanding field. Contributions will be drawn from the main sub-disciplines and from innovative areas of work which have no particular sub-disciplinary allegiances.

For a full list of titles in this series, please visit www.routledge.com/series/SE0514

Place, Diversity and Solidarity

Edited by Stijn Oosterlynck, Nick
Schuermans and Maarten Loopmans

LONDON AND NEW YORK

First published 2017 by Routledge

2 Park Square, Milton Park, Abingdon, Oxfordshire OX14 4RN
52 Vanderbilt Avenue, New York, NY 10017

Routledge is an imprint of the Taylor & Francis Group, an informa business

First issued in paperback 2018

British Library Cataloguing in Publication Data
A catalogue record for this book is available from the British Library

Library of Congress Cataloging in Publication Data
A catalog record for this book has been requested

ISBN: 978-1-138-65497-6 (hbk)
ISBN: 978-0-367-21890-4 (pbk)

Typeset in Times New Roman
by Swales & Willis, Exeter, Devon, UK

Contents

Figures

Tables

Contributors

Harrison Esam Awuh is a doctoral researcher at the Department of Earth and Environmental Sciences at the KU Leuven in Belgium. His research interests include conservation, forced migration, development and poststructuralism. His current work focuses on inequalities associated with conservation-induced population displacement.

Pascal Debruyne worked as a postdoctoral researcher for the DieGem project on diversity, community and solidarity.

David Featherstone is a Senior Lecturer in Human Geography at the School of Geographical and Earth Sciences of the University of Glasgow, Scotland, UK. In focusing on the relations between resistance, space, politics and solidarity, his research contributes to debates on the political geographies of globalisation and internationalism. He is the author of *Solidarity: Hidden Histories and Geographies of Internationalism* (Zed Books) and a member of the editorial collective of *Soundings: a Journal of Politics and Culture*.

Katherine Hankins, Associate Professor of Geography at Georgia State University in Atlanta, Georgia, USA, studies the politics of neighbourhood change. An urban geographer, her work on Atlanta neighbourhoods has examined the ways in which social inequalities around race and class are expressed in and addressed through urban spaces and the interplay between urban spatial change, neighbourhood activism, and other forms of urban politics.

Mervyn Horgan is an Assistant Professor in the Department of Sociology & Anthropology at the University of Guelph, Canada. He is a social theorist and cultural sociologist whose primary research interest is in social sites and scenes where solidarity is passively produced, tenuously assumed or threatened with dissolution, with particular focus on interactions between strangers in diverse societies. This interest is threaded through recent publications in *Journal of Intercultural Studies, Canadian Journal of Sociology* and *Journal of Ethnic and Migration Studies*.

Maarten Loopmans is Associate Professor at the Division of Geography at KU Leuven, Belgium. His work focuses on diversity, social movements, urban policy-making and local participatory processes. He was the guest editor of special issues in *Environment and Planning A* and *Social and Cultural Geography*.

Bruno Meeus is a Senior Lecturer at the Department of Geography at the University of Fribourg, Switzerland. His research interests revolve broadly around the nexus of migration, political economy and urban geography. Between 2012 and 2015, he held a postdoctoral Innoviris grant for a project on the social mobility of New Member States migrants *via* Brussels.

Samuel Nowak is a PhD student in the Department of Geography at the University of California, Los Angeles, USA. An urban geographer, his research is broadly concerned with the relationship between mobility and inequality in the city. His current project examines the politics of urban transit provision, and the ways in which socio-spatial inequalities are reproduced through the securitization of public transit in Los Angeles.

Stijn Oosterlynck is Associate Professor in Urban Sociology at the University of Antwerp, Belgium. His research is concerned with social innovation, welfare state restructuring, urban renewal and new forms of solidarity in diversity.

Nick Schuermans is a postdoctoral researcher at the Centre on Inequalities, Poverty, Social Exclusion and the City of the University of Antwerp and a teaching associate at the geography department of the Vrije Universiteit Brussel, Belgium. His research focuses on segregation, enclave urbanism, encounters and solidarity in diversity. He has been the guest editor of special issues in *Social & Cultural Geography* and *Journal of Housing and the Built Environment*.

Michela Semprebon is Postdoctoral Fellow in the Department of Political Science and Sociology of the University of Bologna, Italy. Until 2015, she was Postdoctoral Fellow at the University of Milan-Bicocca, where she undertook her research on self-building. She obtained a PhD in Urban Sociology at the same university in 2010. Her research interests are: urban policy; social housing; migration and local policy-making; asylum policy and reception; urban conflicts, participation and urban democracy; intergenerational relationships.

Floor Elisabeth Spijkers holds a Master of Science in Cross-Cultural Psychology and is currently a doctoral researcher at the Division of Geography and Tourism at KU Leuven, Belgium. She is also a researcher on the DieGem project on Solidarity in Diversity in which she focuses on leisure and labour settings.

Martina Valsesia has a Bachelor's degree in Sociology from the University of Milan-Bicocca, Italy, where she specialized in territorial studies and public policy. She also received her Associate's degree in Social and Cooperative Housing from Politecnico di Milano. Her professional expertise lies in housing and spatial planning. She has been directly involved in the implementation of cohousing and self-build projects. Valsesia has also acted as a social coordinator in participatory local development projects. She is currently responsible for the planning and management of social housing initiatives as part of the Fondazione Housing Sociale team in Milan, Italy.

Andy Walter, Associate Professor of Geography at the University of West Georgia, USA, is an economic and urban geographer. His research investigates productions of space and place at urban and regional scales from a political economy perspective, including an ongoing study of the capitalist landscapes produced through the sports industry and an examination of Christian social justice activists' commitment to place as a means of addressing poverty. He is also working with colleagues to produce a *People's Guide to Atlanta*.

Mandy de Wilde is a sociologist working in the field of urban studies, citizenship studies and sustainable consumption. She is a postdoctoral researcher at the Environmental Policy Group at Wageningen University, the Netherlands. She has published on local governance, affective citizenship, community participation and the politics of belonging.

Helen F. Wilson is a Senior Lecturer in Human Geography at the University of Manchester, UK. Her work focuses on the geographies of encounter and the politics and possibilities of living with difference. She has published work on tolerance, everyday multiculture, community intervention programmes, and more-than-human ethics. She is co-editor of *Encountering the City: Urban Encounters from Accra to New York* (2016, Routledge).

1 Beyond social capital

Place, diversity and solidarity

*Stijn Oosterlynck, Nick Schuermans and
Maarten Loopmans*

Introduction

Across the world, but notably in the Global North, political elites have called efforts to construct multicultural societies fiascos and attempts to nurture solidarity in diversity failures (Lentin & Titley, 2011). Against the backdrop of renewed geopolitical tensions – which inflate and exploit ethno-cultural, rather than political-economic, cleavages – concerns are raised that both formal mechanisms of redistribution and informal acts of charity, reciprocity and support are challenged by ethnic and cultural diversity (Oosterlynck et al., 2016).

These concerns are echoed in many contemporary social studies on solidarity and diversity, both at the institutional and the interpersonal level. In welfare state studies, claims have been made that multicultural welfare states are challenged both from the side of ethnic-cultural minorities as well as from the majority ethnic-cultural group (Barry, 2001; Kymlicka & Banting, 2006). In conflict studies, a renewed emphasis on institutional racism, originally triggered by unease with slow-paced racial reform in the US, reveals how dealing with ethnic and cultural diversity is connected to power asymmetries and has shown how state institutions tend to aggravate racism and undermine solidarities by privileging majority groups in society (Jones, 2000; Limbert & Bullock, 2005; Gee et al., 2009; Delgado & Stefancic, 2013). At the interpersonal level, the emphasis has long been on the effect of contacts on individual prejudice and racist behaviour (Allport, 1954), neglecting its effects on wider societal structures (Dixon et al., 2005) and merely assuming that individual prejudice is directly related to politics of discrimination and exclusion at the level of society (Jackman & Crane, 1986).

In that respect, Putnam's (2007) seminal analysis of diversity and community-based social capital has been a ground-breaking move towards bridging micro-, meso- and macro-level approaches to solidarity in diversity. Putnam defines social capital as a characteristic of communities, embedded in the social connections between individuals and the norms of reciprocity and trust embedded in them. In his earlier analyses, he discusses social capital to be the glue that keeps society together and allows the development and maintenance of larger-scale systems of solidarity (Putnam, 2001; DeFilippis, 2002). More recently, Putnam has joined the debate on diversity and solidarity from the social capital angle. Partly echoing the pessimistic conclusions in studies on welfare systems or institutional racism, he argues that immigration and ethnic diversity tend to reduce solidarity and social capital in the short to medium run. In fact, his 'constrict theory' presumes that ethnic and cultural diversity, rather than triggering in-group/ out-group division, leads to social isolation and a reduction of both in-group and out-group solidarity. To substantiate this claim, Putnam (2007, p. 149) writes that 'people living in ethnically diverse settings appear to "hunker down" – that is, to pull in like a turtle'. In the long run, however, he remains optimistic that ethnic groups are able to see themselves as members of a shared group with a shared identity and that a broader social solidarity can be constructed out of this, as was the case in previous assimilation processes in American history.

Meanwhile, the extensive literature on social capital in diversity has refined and extended Putnam's initial analysis. While Putnam's (2007) original findings already suggested that individual characteristics such as age, ethnicity, social class and home-ownership provide by far the strongest explanation for differences in social capital between different neighbourhoods, other factors have also been entered into the equation, suggesting that a more contextualized and diversified analysis is needed. Based on a review of 90 studies on the constrict theory, van der Meer and Tolsma (2014) infer, for instance, that there is more evidence for Putnam's hypothesis in the United States than in European countries. In their study of neighbourhoods in the 50 largest Dutch municipalities, Gijsberts et al. (2012) conclude that ethnic diversity has a negative effect on neighbourhood contacts, but not on trust, informal help and voluntary work. Drawing on research in British neighbourhoods, Letki (2008) also found that it is not so much racial diversity but low socio-economic status that undermines social capital.

Even though these studies – and many others – have served to nuance the somewhat blunt policy conclusions that have been drawn from Putnam's original study, they do not fully address the abundant criticisms it has received. Even though Putnam's conceptualization of social capital offers an innovative meso-level perspective to solidarity, scholars argue that the neoclassical economic inspirations and communitarian assumptions underlying his theory considerably reduce the historical depth of the concept of solidarity (Fine, 2001, 2007, 2010; Adkins, 2008; Daly & Silver, 2008). In relation to both the development (Brondolo et al., 2012) and the effects of social capital (Radcliffe, 2004; Portes & Vikstrom, 2011) in contexts of diversity, Putnam remains blind for the effects of power, conflict and racism (Kilson, 2009). This is because Putnam's approach ignores the contextualized practices through which social capital develops, falsely implying social capital works similarly across different spatial and historical contexts. Putnam's take on solidarity is in essence apolitical and aspatial. Social capital is considered the automatic and inevitable outcome of social structures. However, accounts of the political histories and geographies through which solidarities become articulated demonstrate that solidarities, rather than being 'structurally determined', are contingent, malleable and can be struggled over and contested (Mohan & Mohan, 2002; DeFilippis, 2001, 2002; Fine, 2002; Das, 2004; Daly & Silver, 2008; Portes & Vikstrom, 2011; Naughton, 2014; Featherstone, 2016).

Even though Putnam has turned social capital into a global policy buzzword for its individualist, apolitical approach to policy problems like poverty, development, social exclusion, we link its limitations not only to the theoretical and ontological limits of the concept as it is used today, but also to the epistemological and methodological approaches that dominate the social capital literature. While we do not want to deny the merits of trying to understand the interrelationships between ethnic and cultural diversity and social capital through the analysis of large-scale datasets, the difficulties encountered in social capital studies to reach general conclusions bear testimony to the fact that context matters greatly and needs to be taken into account (Naughton, 2014). If we want to develop a deep understanding of how diversity interacts with solidarity, a more place-based and historicizing methodological approach is needed.

As such, the contributions to this volume take different methodological and theoretical routes into the problem of solidarity in diversity. Although we are sympathetic towards the social capital literature's attempt at linking micro-, meso- and macro-social dimensions of solidarity in diversity, the collection of essays in this book adopts an alternative, practice-oriented approach to solidarity. By shifting their analytical focus to the social practices that diverse citizens engage in on a daily basis in specific places, the chapters draw a more diverse and nuanced picture of the possibility of solidarity in diversity. Rather than conducting the kind of survey research that is adopted by most social capital studies, the authors rely on in-depth, qualitative case studies so as to develop a deep understanding of how diversity interacts with solidarity in and through particular spatial and historical contexts. In our view, this presents a more nuanced, or at least more open-ended, view on the prospect of solidarity in contexts of diversity. It enhances our capacity to theorize the politics of solidarity in diversity and the interactions with the places in and through which they unfold. It allows for solidarities to be created 'from below', shaped and initiated by marginalized groups and actors. It also explores the possibilities of conscious interventions to stimulate and strengthen solidarities.

Drawing on a practice-oriented approach to solidarity, this introduction elaborates three new pathways for empirical investigations into the relation between diversity and solidarity. Each of the three following sections presents a critique of the social capital literature, together with an elaboration of the ways in which a practice-oriented approach to solidarity responds to this critique. In each of these three sections, we will situate the contributions to this volume.

The next section starts with a brief overview of sociology's conceptual legacy with regard to solidarity. As such, we explain how Putnam's communitarian take on social capital limits our capacity to seek for alternative *sources of solidarity* in multiethnic and multicultural societies (Featherstone, 2012; Oosterlynck et al., 2016). In the subsequent section, we explain how a practice-oriented approach to solidarity begs the question of place. By engaging with human geography scholarship on the socio-spatial processes that make *relational places* the result, and simultaneously contexts and incubators of new forms of solidarity in diversity, we will reveal the spatio-temporal contingency of solidarities so often overlooked in social capital research (see Amin, 2004; Massey, 2005). In a third and final section, we investigate the *politics of solidarity*, or more specifically, how practices of solidarity shape political identification and group formation and how diversity is an outcome, as much as a determinant of practices of solidarity (Işin, 2002). By emphasizing how practices of solidarity are connected to conflict, power and struggles over citizenship – and in the process shape differences, as much as they are shaped by them – we address the apolitical nature of social capital studies.

From social capital to four sources of solidarity

Putnam's (2007, p. 137) understanding of social capital as the 'social networks and the associated norms of reciprocity and trustworthiness' is strongly related to the communitarian conviction that shared norms and values are the main source of solidarity in society. For Putnam (1993), social solidarities are bound up with the fostering of shared goals through civic engagement. He focuses on the way in which social connections between citizens generate mutual obligations and norms of (generalized) reciprocity that support coordination and cooperation in society. For Putnam, the fabric of the community is, therefore, the main bearer of the norms and values which generate solidarity, not the individual, public policies or state institutions (see Fine, 2001; Radcliffe, 2004; Portes, 2014).

Putnam's work on the decline of community and civic engagement cannot be reduced to communitarianism, but its wide popularity cannot be understood apart from the attraction of the latter. Communitarians also consider the community to be the sphere – alternative to the market and the state – in which social and moral integration is fostered. Morality, according to MacIntyre (1981), is always grounded in the shared beliefs and practices of communities. Individuals, he adds, cannot be the central bearers of morality, as this would lead to a 'society where there is no longer a shared conception of the community's good' (MacIntyre, 1981, p. 215). As such, communitarians argue for members of the community to take care of themselves and their families, but stress that there is also a collective responsibility to take care of community members who are not able to look after themselves. Reciprocity within the community is central to this. Communities develop common goals and their shared values and rules are derived from these common goals. Solidarity is, then, grounded in the common values and shared moral commitments of the community (Oosterlynck & Van Bouchaute, 2013).

Being one of the foundational concepts of the discipline, engagements with solidarity in sociology predate Putnam and much communitarian thinking. Already by the nineteenth century, solidarity was the answer of classical sociologists to the social and political upheaval wrought on society by the associated trends of industrialization, urbanization, migration and secularization. In a quest for the 'glue' that could keep society together, solidarity was defined as the willingness to share and redistribute material and immaterial resources based on feelings of shared fate and group loyalty (Stjernø, 2004, p. 25).

Table 1.1 Four sources of solidarity (adapted from Oosterlynck et al., 2016)

Source	Definition	Example	Authors
Interdependence	Collective benefits of specialization and social differentiation	Social insurance system	- Durkheim on organic solidarity - Spencer on voluntary cooperation - de Tocqueville on civil society - Beck on reflexive individuals - Giddens on active trust - Portes on social capital
Shared norms and values	Moral integration in community of norms and values	Religious organizations	- Durkheim on collective consciousness - Etzioni on reciprocity in the community - Putnam on social capital
Struggle	Incompatibility of interests and visions between two or more social groups	Labour class movement	- Marx on the unity of the working class - Weber on class and status groups - Bourdieu on social capital
Encounter	Informal social interaction with strangers	Living with strangers in public space	- Simmel on sociation - Chicago School of urban sociology

Going through the canon of sociological work on solidarity, it is obvious that both Putnam and the communitarians have a particular perspective on the grounds of solidarity. Elsewhere, we have argued that classical sociologists have identified at least four distinctive sources of solidarity (Table 1.1; see Oosterlynck et al., 2016). Each of these four sources reflects fundamentally different analytical as well as normative positions on the grounds of social order and cohesion in society (Silver, 1994). Putnam's understanding of solidarity in terms of social capital mainly mobilizes one of these four sources, namely shared norms and values. This is linked to the communitarian belief that solidarity is only possible through moral integration into a community with common commitments. Obviously, shared norms and values can be a strong source of solidarity. Meeus's contribution to this volume provides a good example. In Brussels, he explores how a Romanian migrant church is a hub where different pre-existing forms of social capital that draw on shared biographies, shared religious practices, shared regional origins and shared norms and values are tied together so that situational solidarity is nurtured.

By looking at solidarity, rather than social capital, this volume aims to depart from shared norms and values as the sole route into solidarity. Even though this chapter does not allow for an elaborate discussion of the three remaining sources of solidarity (see Oosterlynck et al., 2016), we will introduce them here briefly. We will also make visible how the new practices of place-based solidarity on which the contributions to this volume focus do not only mobilize shared norms and values, but also one or more of the three other sources of solidarity.

First, interdependence generates solidarity through an awareness of the collective benefits of specialization and social differentiation. Informed by the liberal belief that society is (or should be) built through voluntary association between free individuals, this source of solidarity is driven by the conviction that cooperation, trust and mutual reliance are required to reap the benefits of a specialized division of labour or socio-cultural differentiation (e.g. Durkheim, 1893). In their chapter on self-building housing experiments in the north of Italy, Semprebon and Valsesia show, for instance, how a learning process associated with a shared experience of interdependence can bring about solidarity. They also explain, however, how the same mutual interdependence, which underlies also an individual objective for self-builders to access decent housing for themselves, can create tensions and conflicts, which often establish themselves upon the identitarian boundaries of ethnicity and culture. Interdependence as a source of solidarity is also strongly present in the chapter by Awuh and Spijkers. Building upon Allport's (1954) contact hypothesis, they explain how the pursuit of common goals and inter-group cooperation during a football tournament and the digging of latrines in south-east Cameroon improves the relationships between the local Baka and Bantu.

Second, struggle implies a situation of incompatible interests and visions. This source of solidarity is informed by a conflict perspective on society (often associated with left-wing ideologies such as socialism), in which power plays an important role in determining societal outcomes and social order. Feelings of shared fate and group loyalty among one group are, then, generated by mobilizing against another group (e.g. Weber, 1922). Drawing on a case study of the Community Leadership Network, an international training organization with more than 50 branches across Europe and North America, Wilson's chapter in this volume points at the importance of emotional ties, embodied experiences and affective attachments in international struggles for social justice. In Meeus's chapter, the shared struggle against flexible labour contracts, social fraud and privatization policies is an important source of solidarity between Belgians and Romanians working in train stations in Brussels. Bridging the divides between workers with different ethnicities and/or nationalities, they fight together for better labour conditions, better salaries and better public services. In the context of the migrant church, Meeus also illustrates, however, that a shared sense of ethnic 'misrecognition' in Brussels, feeds ethnicized meritocratic values as well.

Lastly, encounter can generate solidarity through informal social interaction with strangers. Through interpersonal interaction, however superficial and brief, an informal and largely implicit and tacit social order emerges through which social life is regulated and which enables people to live together with strangers (e.g. Simmel, 1950; Goffman, 1983). In his discussion of the interactions between gentrifiers and roomers in Toronto, Mervyn Horgan ties most closely to the micro-sociological tradition of encounter. In bridging Goffman's and Simmel's work to contemporary recognition scholarship, he discusses how the mundane practices of mutual recognition become a minimal precondition for what he calls the 'soft solidarity' between strangers implicitly conveyed through the intersubjective accomplishment of ordinary everyday urban life. Solidarity, for Horgan, is expressed in the urban interaction order of mutual indifference ('civil inattention') identified by Goffman, which makes life between strangers possible. The existence of this order, according to Horgan, points at the mutual recognition of both the presence and the difference of strangers in a particular place, a fundamental precondition for soft solidarities to develop.

Obviously, in real-life situations, different sources of solidarity tend to appear side by side. In their account of intentional neighbours in Atlanta – middle-class people who opt to relocate to poor neighbourhoods deliberately – Walter, Hankins and Nowak do not only explain, for instance, how important faith is to become advocates of social and spatial justice, but also how living with the poor allows them to tap into other sources of solidarity, from the shared struggle against institutional desertification to a shared set of strong family values. The chapter by Schuermans and Debruyne reveals a similar mix of sources of solidarity. In their research on a primary and pre-primary school in the university town of Leuven (Belgium), they point at shared norms and values, struggle and encounter as crucial sources of solidarity. By drawing on a wider set of sources of solidarity, the contributions to this volume, thus, move beyond the monopoly of shared norms and values in the social capital literature.

Placing solidarity: encounters in relational places

Putnam (2007) locates solidarity in a community of shared norms and values that is nurtured through the transfer of cultural and normative frameworks from one generation to another. When these shared norms and values are challenged, he suggests that the construction of a new and more inclusive 'we' through repeated social interactions over long periods of time will generate new – and more encompassing – solidarities. As such, the strong boundaries of the nation-state are instrumental in the development of solidarity, as they help to contain social relations, economic interactions and political dynamics within supposedly culturally homogeneous populations. The underlying assumption is that feelings of shared fate and group loyalty are rooted in a strong feeling of shared territory. As such, Putnam's (2007) concept of solidarity is predicated on the territorial containment and historical continuity of the nation-state.

In this volume, we argue for a focus on forms of solidarity in diversity that develop around the joint appropriation and envisaged common future of specific places. By shifting our spatio-temporal perspective from the territoriality and the *longue durée* of the nation-state to the here and now of the practices of ethnically and culturally diverse people in diverse places, we aim to provide an alternative research agenda for the 'loss of community' narrative that recurs in so many studies on social capital and the welfare state. As such, we focus on the places where people of different backgrounds encounter each other.

Obviously, it cannot be assumed that encounters in places marked by diversity will generate solidarity automatically. In fact, a lot of empirical evidence suggests that many engagements across difference confirm stereotypes and reproduce practices of exclusion or disengagement (Valentine, 2008; Schuermans, 2016). Similar nuances can be found in several chapters in this volume. Drawing on ethnographic fieldwork in Amsterdam, De Wilde describes how volunteers involved in a community participation program engage in a 'civility towards diversity' (Wessendorf, 2014, p. 394) – a fine balance between building positive relations and keeping a distance – but the bulk of the activities and interactions they engage in are based on commonalities rather than differences. The chapter by Horgan raises similar concerns. Arguing that civil inattention and mutual indifference are central to soft solidarity, he also warns that structural inequalities are reinforced in public space encounters when indifference is non-mutual. The interactional form of indifference celebrated by Goffman (1983), Horgan writes, reproduces the city as an exclusionary social form when characterized by asymmetry.

As such, these studies invalidate a simplistic account of encounter. For encounter to effectuate more fundamental changes in intercultural practices and values, more is needed than superficial contacts with strangers on buses, trains, streets, squares or parks (Matejskova & Leitner, 2011; Schuermans, 2017). Only more prolonged and intense engagement, Amin (2002, p. 970) argues, 'disrupts easy labelling of the stranger as enemy and initiates new attachments'. Recurring interactions in these places 'are moments of cultural destabilisation', he insists, 'offering individuals the chance to break out of fixed relations and fixed notions, and through this, to learn to become different through new patterns of social interaction' (Amin, 2002, p. 970).

Many contributions to this volume elaborate upon Amin's thesis. They explore the relation between the conditions of encounter, the concrete practices that people engage in at such places of encounter and the solidarities that are or are not shaped by it. Meeus, for instance, calls Brussels train stations pedagogical environments, both for Belgian members of trade unions and for non-unionized Romanian workers, because regular interactions over the course of several years encouraged both groups to move from a meritocratic logic centring around competition to a more cooperative logic marked by inter-ethnic solidarity. Focusing on the spaces in which friendships and solidarities across language, race, ethnicity, age, gender and sexuality are built, Wilson underlines the importance of annual meetings and regular training events of the Community Leadership Network. While she recognizes the possibilities Facebook or Skype provide to communicate over long distances, she emphasizes that both the formal trainings around prejudice and discrimination and the more informal dinners, breakfasts or trips to the supermarket are crucial to establish the affective bonds that sustain the voluntary practices and social activism of the members.

In their action research on the outskirts of the Dja Reserve in south-east Cameroon, Awuh and Spijkers demonstrate how prolonged and profound engagement in mixed teams can challenge the stereotypes people of different ethnic backgrounds have about each other, even if they have been living segregated lives in the same villages for years. By engaging with the notion of 'spatial solidarity', Walter, Hankins and Nowak underline that Christian community development – and the solidarities it sustains – cannot be done effectively over long distances and requires relocation. Even though they always have the option to move out, the spatial propinquity of intentional neighbouring allows relocators to experience and understand the challenges poor people face. Through daily interactions along lines of race and class, it becomes difficult to maintain neoliberal accounts of poverty which blame the poor for their predicament, ultimately paving the way for more progressive perspectives on poverty and privilege and advanced relationships of solidarity.

A more precise understanding of the complex interactions between encounter and solidarities requires a rethinking of the nature of places of encounter and, more specifically, a shift from a territorial towards a relational understanding of place (Massey, 2004, 2005; Amin, 2004; Routledge & Cumbers, 2009; Featherstone, 2012). Relational places are not characterized by boundaries and fixity, but by connections and fluidity (Massey, 2005). In Doreen Massey's (1991) now classic description of Kilburn High Road, a multicultural shopping street in London, she asserts that the qualities of the place as a site of co-existence and encounter are not so much determined by its internal history, but upon the particular constellation of economic relations, social networks and cultural webs which link that street with places all over the world.

If we acknowledge that the places where we spend our everyday lives are the product of relations that stretch way beyond them, then the prospect – or maybe even the moral obligation – to develop solidarity across borders needs to be considered (Darling, 2010). While it is often assumed that proximity is a prerequisite for solidary behaviour and that distance leads to indifference (see Smith, 1998; Barnett, 2005), a relational understanding of place does not only call for solidarity with people we feel spatially and socially close to, but also with strangers in faraway places we are economically, socially or culturally connected with (Darling, 2009). In the framework of Amin (2004), such solidarity can be called solidarity in connectivity.

Because of economic relations, cultural connections and social networks of immigrants, relational places are necessarily marked by inescapable heterogeneity and hybridity, rather than by cultural sameness and intra-territorial homogeneity (Massey, 2005). This implies that solidarity developed in relational places cannot only be based on a recognition of linkages with other places, but also on an appreciation of the social, cultural and economic differences within them. Following Amin (2004), such a solidarity across different communities, cultures and identities in spatial proximity can be called solidarity in propinquity.

Rather than presenting a relational understanding of place as a normative ideal in the quest for solidarity in diversity, this volume deals with empirical explorations of how propinquity and connectivity are enacted in everyday struggles and engagements (see Barnett & Land, 2007; Featherstone, 2012). Indeed, much of the literature on relational places is of a theoretical and normative character and avoids empirical discussion of the mundane processes through which these solidarities of propinquity and connectivity are established (Jones, 2009). Responding to the critique that much of the literature on relational places lacks an empirical basis, the different contributors to this volume focus on the way in which (places of) encounters across lines of class, race, culture or ethnicity nurture solidarities with close and distant others by reconfiguring the definition of communities and places. In schools, parks, factories, buses, offices, sports fields or neighbourhood centres, these forms of solidarity are not rooted in philosophical discussions about the nature of space, but in everyday interactions, negotiations and confrontations across difference (Valentine, 2008; Matejskova & Leitner, 2011; Wilson, 2016; Schuermans, 2016).

Various contributions to this volume bear witness to solidarities in propinquity, as well as connectivity. Drawing on her own participation in the Community Leadership Network, Wilson infers, for instance, how interpersonal relations, attachments and friendships are able to connect place-based struggles and spaces of activism in radically different local and national contexts. In their empirical research in a Belgian primary and pre-primary school, Schuermans and Debruyne also indicate how issues of economic redistribution, cultural recognition and political participation are not only addressed within the school walls, but also in a much wider network of educational and non-educational partners at the neighbourhood, city and region-wide scales. From a more theoretical perspective, Horgan is interested in socio-spatial justice that is not rooted in 'regressive territorialism', as he calls it. While his minimalist conception of solidarity is spatially circumscribed because it is based on physical co-presence, it is not territorially confined in the way nation-based solidarities are.

Politicizing solidarity: solidarity and citizenship practices

Drawing on Massey's (2005) relational account of 'throwntogetherness', David Featherstone, in his afterword to this volume, emphasizes that it is 'crucial to note how spaces of organizing and encounter have shaped and been shaped by solidarities which articulate new ways of relating and struggling against unequal ways of generating place'. Engaging with this place-making character of solidarity is central to thinking solidarities politically. In Featherstone's words, it is important to understand 'how solidarities can entrench, challenge and rework relations of power within particular places'.

Engin Işin's (2002) genealogical writings on citizenship provide a fruitful inroad to conceptualize the political nature of solidarity. In a virulent critique of Weber's orientalist reading of citizenship, Işin explains how citizenship, as forms of solidarity that transcend bonds of kinship and rest on alternative sources than (imagined) genetic closeness, has falsely been considered an outcome of socio-spatial unification (also by communitarian interpretations such as Putnam's). Instead, Işin's 'politicized' solidarities emerge from diverse relational strategies or practices which lead to differentiation as much as to unification. Addressing solidarity from the angle of relational practices/strategies instead of from the angle of sources, Işin analyses how solidarity is enacted and is essentially a process, permanently in the making.

In his conceptualization of the city as a 'difference machine', Işin (2002) analyses how practices of solidarity generate new forms of identification and divisions between citizens and their subaltern others (strangers, outsiders and aliens).

As elaborated in the chapter by De Wilde, the boundaries and the hierarchies between these different groups are created and moulded through solidaristic, agonistic and alienating practices of citizenship. The four sources of solidarity outlined above can be tied to this creation of difference. First, the community of citizens is determined by solidaristic practices, such as sociation, identification and affiliation. Citizen-to-citizen solidarities develop on the basis of the recognition of shared norms and values. Second, the relation between citizens and strangers is established through solidaristic as well as agonistic practices such as conflict, competition, domination and resistance. In the context of citizen–stranger relations, oppositional solidarities can develop based on struggle over claims to the same social goods. Third, outsiders are constituted as 'neither belonging to the group nor interacting with it, but belonging to and necessary for the city in which citizens and strangers associated' (Işin, 2002, p. 31). Citizen–outsider relations are often a combination of agonistic and alienating strategies, like exclusion, estrangement, oppression, expulsion. The solidarities that can develop in such relationships, are based on a conscience of mutual interdependence. Finally, aliens are defined by alienation strategies only. Whereas the three former groups are constituted as immanent in and constitutive of a place, the alien is of a more transient character. They are not considered necessary, and engagement with aliens is of a temporary, fleeting character. At best, they are addressed with the blasé indifference emphasized in the literature on encounter (Simmel, 1950), at worst, with exclusion, expulsion or annihilation.

Işin's work politicizes solidarity, in that it reveals how the various sources of solidarity can be linked to gradually expanding definitions of citizenry, and associated with different combinations of citizenship practices oriented to the development of identities, boundaries and distinctions, inclusion and exclusion. Işin also emphasizes how space is co-constitutive of these solidarity politics: 'In their encounters and interactions that constitute space (configurations of objects expressed in boundaries, buildings, zones, insides and outsides) agents simultaneously form themselves and others as groups (expressed in citizen–citizens, citizen–strangers and citizens–outsiders, citizens–aliens logics)' (Işin, 2002, p. 49). Space is not a neutral background to these politics, but is a fundamental strategic property; it can be a source for identification and the construction of common reference frames (Martin, 2003, 2013; Osler & Starkey, 2005); it can become a focus of struggle (Castells, 1983; Mayer, 2003; Loopmans & Dirckx, 2012; Loopmans et al., 2012), or it can be a locale where relations of interdependence are forged and made explicit (Simone, 2004, 2010; Routledge & Cumbers, 2009; Nicholls, 2009; Featherstone, 2012; Miller & Nicholls, 2013).

In this volume, we analyse how the relational context of superdiversity, where ethnic and cultural diversity characterizes relations of propinquity as well as connectivity, constitute a particular 'spatial differentiation machine'. By investigating solidarity, the different contributors to this book are less interested in citizenship as an individually owned status to support claims to particular rights, but focus on citizenship as practices supporting relations of solidarity. This implies that we explicitly look for moments in which actors make issues of economic redistribution, cultural recognition and political representation visible and the subject of debate (see Fraser, 1995, 2010; Oosterlynck et al., 2016). In paying close attention to how and to what extent the observed practices of solidarity intersect with processes of social exclusion and are able to redress various social injustices, we aim to better understand the progressive potential of place-based forms of solidarity in diversity in relational settings.

The 'inventive' character of solidarities (Featherstone, 2012) raises the question how solidaristic, agonistic and alienating strategies and practices are played out through and in the spatial structures of diverse relational places, and what kind of solidarities are constructed through them. Are the group subjectivities developed in relational places territorially bound, or do they stretch far beyond territorial boundaries (Massey, 2004, 2005)? Are shared frames of identification and norms and values local, or do they obtain a global span (Routledge & Cumbers, 2009; Pierce et al., 2011)? Are networks of interdependence, embedded in the agonistic and alienating interaction among strangers and outsiders in relational places, to be seen as wide-ranging infrastructures of interdependence-based solidarities (Simone, 2004)?

These questions are taken up in several chapters of this volume. Arguing that soft solidarities nurtured in public space encounters marked by mutual indifference are necessary for the achievement of social justice, Horgan demonstrates, for instance, how alienating and solidarizing strategies go together to challenge and reinforce the boundaries between new homeowners and the less privileged 'roomers' they refuse to see as neighbours in their gentrifying neighbourhood. Along similar lines, Walter, Hankins and Nowak show how practices of intentional neighbouring result in processes of identification that shift the subject boundaries from aliens and outsiders to stranger (and hence from charity to citizenship), so that interlocking structures of oppression and exploitation are being challenged. In her ethnographic work on the Community Leadership Network, Wilson also demonstrates how intimate encounters work to disrupt assumptions of sameness and difference.

As David Featherstone emphasizes in his afterword to this volume, the constitution of political identities and solidarities through everyday relational interactions does not happen in a power vacuum. Işin identifies citizenship practices as routinized activities that have institutional characteristics. They are overdetermined by the assemblage of certain groups with institutional networks of state and para-state organizations which underpin structures of privilege and marginalization. Through the deployment of 'street level bureaucrats' and formally independent professions (Scott, 1998; Cruikshank, 1999; Rose, 1999; Loopmans, 2006), citizenship as solidarity is constituted through the entanglement of state and legal institutions with everyday practices (Staeheli et al., 2012). Public institutions such as the education system or the labour market, and the concrete professionals working in them, play an active role in moral integration and the shaping of group identities. As such, it can be questioned, for instance, whether interventions of street level bureaucrats and professionals limit or expand the scope of particular struggles (Sites, 2012). It is also crucial to find out whether their entanglement with formal institutions of solidarity brings about 'relational incubators' for extended solidarities or citizenship subjectivities within place boundaries (see Uitermark et al., 2012).

To answer these questions, the different contributors to this volume do not only look at the role of relational places as 'machines' for citizenship strategies and technologies. By studying the practices through which relational place-based forms of interaction create citizenship or group subjectivities which support forms of sharing material and non-material resources among cultural and ethnic increasingly diverse populations, it becomes clear that mediating actors often play a crucial role. In Semprebon and Valsesia's chapter, the coordinators of the building project are essential in steering the interactions between people. Similarly, in the chapter by Awuh and Spijkers, the action-researcher is the one who facilitates the activities (football competition and latrine building) that stimulate reciprocal learning practices. In De Wilde's case study in Amsterdam, policy practitioners working for a community participation programme take up a prominent role. And in the case described by Schuermans and Debruyne, current and former school principals are crucial in deciding upon the place and character of interactions between teachers and children, among children and between teachers and the neighbourhood.

Several chapters of this volume also emphasize the role of institutions in affecting the interactional practices of diverse individuals to achieve solidarity across difference. In the chapter by Walter, Hankins and Nowak, as much as in the one by Meeus, institution of faith, and the practices that are exercised in the name of faith, are crucial in stimulating the development of solidarities. The institutions of the trade union (Meeus), the school (Schuermans and Debruyne), the Community Leadership Network (Wilson), the self-building cooperative and the building society (Semprebon and Valsesia) similarly stimulate individual practices of sharing, collaboration and interdependence that lead to the reproduction of solidarities. Moreover, as is underlined in Meeus's account of the church and the trade union, but also by Schuermans and Debruyne in the case of a school, the development and expansion of cross-difference solidarities also fundamentally alters the structures of the institution. Hence, it is clear that solidarities and institutions co-constitute each other in a mutually reinforcing way.

Conclusion

Drawing on insights from the disciplines of sociology, geography and political science, we have argued in this introductory chapter that micro- and macro-level analyses of solidarity in diversity need to be bridged. Our central claim is that Putnam's conceptualization of social capital provides an interesting departure point to start thinking about the interrelations between the imagined community of the nation-state and the glue that keeps a particular place together, but that his thinking is impoverished by maintaining a communitarian focus on shared norms and values as a source of solidarity and by ignoring the relational places in which solidarity is nurtured, together with the political practices through which solidarity unfolds. As such, it is our contention that Putnam's understanding of solidarity can be enriched by 1) considering a wider range of sources of solidarity; 2) by situating solidarity in relational places; and 3) by politicizing solidarity through citizenship practices.

First, we have indicated how classical social theorists have identified four sources of solidarity which remain relevant today (interdependence, shared norms and values, struggle and encounter). Issues of inclusion and exclusion are central to each of these four sources of solidarity. Interdependence requires us to make explicit whom we are interdependent with and whom we are not dependent of; solidarity in struggle involves the definition of an enemy outsider against which to unite; shared norms and values as source of solidarity excludes those who do not conform to the *mores* of the community; encounters can only be a source of solidarity for those who do not circumvent or curtail them beforehand. As with social capital (see Portes, 1998; Daly & Silver, 2008), practices of solidarity are, thus, productive of boundaries, irrespective of the sources they tap into. This question of belonging does not need to be solved on the basis of similarities in skin colour, ethnicity or culture, however, but is the contingent and temporary outcome of a spatio-temporal politics of alliances and solidarity (Agustín & Jørgensen, 2016).

The fact that solidarity is characterized by social boundedness does not necessarily imply it is also marked by spatial boundedness, however. In fact, we claim that solidarities grounded in the territorial boundedness of nation-states are complemented and enriched today with solidarities developing in the entirely different spatio-temporal frame of the relational places in which encounters across ethnic and cultural boundaries occur (see Oosterlynck et al., 2016). The solidarities we are looking for in this volume do not presuppose integration into a predefined community, nor the necessity of historical time to build up social capital between diverse citizens, but require a willingness to negotiate the diversity of people, practices, claims and networks that affect a particular place. The recognition that neighbourhoods, cities and countries are made through cultural, social and economic circuits that reach far beyond their territorial boundaries, implies that such solidarities are place-based, but not place-bound (see Amin, 2004).

Often, it is argued that place-based studies of solidarity in diversity ignore deeper political, economic and social structures which produce and reproduce ethnic disadvantages, racist ideologies and socio-economic inequalities (e.g. Kymlicka, 2016). The various chapters in this book bear witness, however, to the fact that solidarities nurtured through encounters in relational places are not limited to 'light' forms of solidarity such as conviviality (Wessendorf, 2014) or indifference to difference (Amin, 2012). After all, they reveal how relational places are 'citizenship machines' (Işin, 2002) which are shaped by – and shape – political strategies and solidarity practices beyond the local (Featherstone, 2012). In this way, solidarities that develop around very specific issues in particular places are able to question and to challenge state policies relating to social benefits, the educational system or the labour market and the structural mechanisms of exclusion embedded in them.

References

Adkins, L. (2008). Social capital put to the test. *Sociology Compass*, *2*(4), 1209–1227.

Agustín, G. O., & Jørgensen, M. B. (2016). Against pessimism: A time and space for solidarity. In O. G. Agustín & M. B. Jørgensen (Eds.), *Solidarity without borders: Gramscian perspectives and civil society alliances* (pp. 3–20). London: Pluto Press.

Allport, G. (1954). *The nature of prejudice*. Reading, MA: Addison-Wesley.

Amin, A. (2002). Ethnicity and the multicultural city: Living with diversity. *Environment and Planning A*, *34*, 959–980.

Amin, A. (2004). Regions unbound: Towards a new politics of place. *Geografiska Annaler B*, *86*, 33–44.

Amin, A. (2012). *Land of strangers*. Cambridge: Polity Press.

Barnett, C. (2005). Ways of relating: Hospitality and the acknowledgement of otherness. *Progress in Human Geography*, *29*, 5–21.

Barnett, C., & Land, D. (2007). Geographies of generosity: Beyond the 'moral turn'. *Geoforum*, *38*, 1065–1075.

Barry, B. (2001). *Culture and equality. An egalitarian critique of multiculturalism*. Cambridge: Polity Press.

Brondolo, E., Libretti, M., Riviera, L., & Walsemann, K. M. (2012). Racism and social capital: The implications for social and physical wellbeing. *Journal of Social Issues*, *68*, 358–384.

Castells, M. (1983). *The city and the grassroots: A cross-cultural theory of urban social movements*. Oakland: University of California Press.

Cruikshank, B. (1999). *The will to empower: Democratic citizens and other subjects*. New York: Cornell University Press.

Daly, M., & Silver, H. (2008). Social exclusion and social capital: A comparison and critique. *Theory and Society*, *37*(6), 537–566.

Darling, J. (2009). Thinking beyond place: The responsibilities of a relational spatial politics. *Geography Compass*, *3*, 1938–1954.

Darling, J. (2010). A city of sanctuary: The relational re-imagining of Sheffield's asylum politics. *Transactions of the Institute of British Geographers*, *35*, 125–140.

Das, R. J. (2004). Social capital and poverty of the wage-labour class: Problems with the social capital theory. *Transactions of the Institute of British Geographers*, *29*(1), 27–45.

DeFilippis, J. (2001). The myth of social capital in community development. *Housing Policy Debate*, *12*(4), 781–806.

DeFilippis, J. (2002). Symposium on social capital: An introduction. *Antipode*, *34*(4), 790–795.

Delgado, R., & Stefancic, J. (2013). Discerning critical moments. In M. Lynn & A. Dixson (Eds.), *Handbook of critical race theory in education* (pp. 23–33). New York: Routledge.

Dixon, J., Durrheim, K., & Tredoux, C. (2005). Beyond the optimal contact strategy: A reality check for the contact hypothesis. *American Psychologist*, *60*(7), 697–711.

Durkheim, E. (1893). *The division of labour in society*. London: Macmillan.

Featherstone, D. (2012). *Solidarity: Hidden histories and geographies of internationalism*. London: Zed Books.

Featherstone, D. (2016). Politicising the crisis: the Southern Question, uneven geographies and the construction of solidarity. In O. G. Agustín & M. B. Jørgensen (Eds.), *Solidarity without borders: Gramscian perspectives and civil society alliances* (pp. 169–185). London: Pluto Press.

Fine, B. (2001). *Social capital versus social theory: Political economy and social science at the turn of the century*. London: Routledge.

Fine, B. (2002). They f** k you up those social capitalists. *Antipode, 34*(4), 796–799.

Fine, B. (2007). Eleven hypotheses on the conceptual history of social capital: A response to James Farr. *Political Theory, 35*(1), 47–53.

Fine, B. (2010). *Theories of social capital: Researchers behaving badly*. London: Pluto Press.

Fraser, N. (1995). From redistribution to recognition? Dilemmas of justice in a 'postsocialist' age. *New Left Review, 212*, 68–93.

Fraser, N. (2010). *Marketization, social protection, emancipation: toward a neopolanyian conception of capitalist crisis*. Paper presented at the Seminar Series on the Institutions that Manage Violent Conflict, Columbus, November 18.

Gee, G. C., Ro, A., Shariff-Marco, S., & Chae, D. (2009). Racial discrimination and health among Asian Americans: Evidence, assessment, and directions for future research. *Epidemiologic Reviews, 31*(1), 130–151.

Gijsberts, M., van der Meer, T., & Dagevos, J. (2012). 'Hunkering down' in multi-ethnic neighbourhoods? The effects of ethnic diversity on dimensions of social cohesion. *European Sociological Review, 28*(4), 527–537.

Goffman, E. (1983). The interaction order. *American Sociological Review, 48*, 1–17.

Işin, E. (2002). *Being political. Genealogies of citizenship*. London and Minneapolis: University of Minnesota Press.

Jackman, M. R., & Crane, M. (1986). 'Some of my best friends are black. . .': Interracial friendship and whites' racial attitudes. *Public Opinion Quarterly, 50*(4), 459–486.

Jones, C. P. (2000). Levels of racism: A theoretic framework and a gardener's tale. *American Journal of Public Health, 90*(8), 1212–1215.

Jones, M. (2009). Phase space: Geography, relational thinking, and beyond. *Progress in Human Geography, 33*(4), 487–506.

Kilson, M. (2009). Thinking about Robert Putnam's analysis of diversity. *Du Bois Review: Social Science Research on Race, 6*(02), 293–308.

Kymlicka, W. (2016). Rejoinder from sociability to solidarity: Reply to commentators. *Comparative Migration Studies*, DOI: 10.1186/s40878-016-0030-2.

Kymlicka, W., & Banting, K. (2006). Immigration, multiculturalism, and the welfare state. *Ethics & International Affairs, 20*, 281–304.

Lentin, A., & Titley, G. (2011). *The crises of multiculturalism*. London: Zed Books.

Letki, N. (2008). Does diversity erode social cohesion? Social capital and race in British neighbourhoods. *Political Studies, 56*(1), 99–126.

Limbert, W. M., & Bullock, H. E. (2005). 'Playing the fool': US welfare policy from a critical race perspective. *Feminism & Psychology, 15*(3), 253–274.

Loopmans, M. (2006). From residents to neighbours: The making of active citizens in Antwerp, Belgium. In J. Duyvendak, T. Knijn & M. Kremer (Eds.), *Professionals between people and policy: Transformations in care and welfare in Europe* (pp. 109–121). Amsterdam: Amsterdam University Press.

Loopmans, M., Cowell, G., & Oosterlynck, S. (2012). Photography, public pedagogy and the politics of place-making in post-industrial areas. *Social & Cultural Geography, 13*(7), 699–718.

Loopmans, M., & Dirckx, T. (2012). Neoliberal urban movements? A geography of conflict and mobilisation over urban renaissance in Antwerp, Belgium. In T. Taşan-Kok & M. Raco (Eds.), *Contradictions of neoliberal planning* (pp. 99–116). Amsterdam: Springer.

MacIntyre, A. (1981). *After virtue: A study in moral theory*. London: Duckworth.

Martin, D. G. (2003). 'Place-framing' as place-making: Constituting a neighborhood for organizing and activism. *Annals of the Association of American Geographers, 93*(3), 730–750.

Martin, D. G. (2013). Place frames: Analysing practice and production of place in contentious politics. In W. Nicholls, B. Miller & J. Beaumont (Eds.), *Spaces of contention: Spatialities and social movements* (pp. 85–99). Aldershot: Ashgate.

Massey, D. (1991). A global sense of place. *Marxism Today, 35*, 24–29.

Massey, D. (2004). Geographies of responsibility. *Geografiska Annaler B, 86*, 5–18.

Massey, D. (2005). *For space*. London: SAGE.

Matejskova, T., & Leitner, H. (2011). Urban encounters with difference: The contact hypothesis and immigrant integration projects in eastern Berlin. *Social & Cultural Geography, 12*, 717–741.

Mayer, M. (2003). The onward sweep of social capital: Causes and consequences for understanding cities, communities and urban movements. *International Journal of Urban and Regional Research, 27*(1), 110–132.

Miller, B., & Nicholls, W. (2013). Social movements in urban society: The city as a space of politicization. *Urban Geography, 34*(4), 452–473.

Mohan, G., & Mohan, J. (2002). Placing social capital. *Progress in Human Geography, 26*(2), 191–210.

Naughton, L. (2014). Geographical narratives of social capital. Telling different stories about the socio-economy with context, space, place, power and agency. *Progress in Human Geography, 38*(1), 3–21.

Nicholls, W. (2009). Place, networks, space: Theorising the geographies of social movements. *Transactions of the Institute of British Geographers, 34*(1), 78–93.

Oosterlynck, S., Loopmans, M., Schuermans, N., Vandenabeele, J., & Zemni, S. (2016). Putting flesh to the bone: Looking for solidarity in diversity, here and now. *Ethnic and Racial Studies, 39*, 764–782.

Oosterlynck, S., & Van Bouchaute, B. (2013). Een gemeenschap van gedeelde normen en waarden: doodlopende weg naar solidariteit? [A community of shared norms and values: dead-end-street to solidarity?], *Ophouwwerk Brussel, 208*, 35–44.

Osler, A., & Starkey, H. (2005). *Changing citizenship: Democracy and inclusion in education*. Maidenhead: Open University Press.

Pierce, J., Martin, D. G., & Murphy, J. T. (2011). Relational place-making: The networked politics of place. *Transactions of the Institute of British Geographers, 36*(1), 54–70.

Portes, A. (1998). Social capital: Its origins and applications in modern sociology. *Annual Review of Sociology, 24*, 1–24.

Portes, A. (2014). Downsides of social capital. *Proceedings of the National Academy of Sciences, 111*(52), 18407–18408.

Portes, A., & Vikstrom, E. (2011). Diversity, social capital, and cohesion. *Annual Review of Sociology, 37*, 461–479.

Putnam, R. D. (1993). The prosperous community: Social capital and public life. *The American Prospect, 13*, 35–42.

Putnam, R. D. (2001). *Bowling alone: The collapse and revival of American community*. New York: Simon & Schuster.

Putnam, R. D. (2007). E pluribus unum: Diversity and community in the twenty-first century. The 2006 Johan Skytte Prize Lecture. *Scandinavian Political Studies, 30*, 137–174.

Radcliffe, S. A. (2004). Geography of development: Development, civil society and inequality – social capital is (almost) dead? *Progress in Human Geography, 28*, 517–527.

Rose, N. (1999). *Powers of freedom: Reframing political thought.* Cambridge: Cambridge University Press.

Routledge, P., & Cumbers, A. (2009). *Global justice networks: Geographies of transnational solidarity.* Manchester: Manchester University Press.

Schuermans, N. (2016). Enclave urbanism as telescopic urbanism? Encounters of middle class whites in Cape Town. *Cities, 59,* 183–192.

Schuermans, N. (2017). White, middle-class South Africans moving through Cape Town: Mobile encounters with strangers. *Social & Cultural Geography, 18*(1), 34–52.

Scott, J. C. (1998). *Seeing like a state: How certain schemes to improve the human condition have failed.* New Haven, CT: Yale University Press.

Simmel, G. (1950). The metropolis and mental life. In K. Wolff (Ed.), *The sociology of Georg Simmel* (pp. 409–424). New York: Free Press.

Simone, A. (2004). People as infrastructure: Intersecting fragments in Johannesburg. *Public Culture, 16*(3), 407–429.

Simone, A. (2010). *City life from Jakarta to Dakar: Movements at the crossroads.* London: Routledge.

Silver, H. (1994). Social exclusion and social solidarity: Three paradigms. *International Labour Review, 133,* 531–578.

Sites, W. (2012). God from the machine? Urban movements meet machine politics in neoliberal Chicago. *Environment and Planning A, 44*(11), 2574–2590.

Smith, D. M. (1998). How far should we care? On the spatial scope of beneficence. *Progress in Human Geography, 22,* 15–38.

Staeheli, L. A., Ehrkamp, P., Leitner, H., & Nagel, C. R. (2012). Dreaming the ordinary daily life and the complex geographies of citizenship. *Progress in Human Geography, 36*(5), 628–644.

Stjernø, S. (2004). *Solidarity in Europe.* Cambridge: Cambridge University Press.

Uitermark, J., Nicholls, W., & Loopmans, M. (2012). Cities and social movements: Theorizing beyond the right to the city. *Environment and Planning A, 44,* 2546–2554.

Valentine, G. (2008). Living with difference: Reflections on geographies of encounter. *Progress in Human Geography, 32,* 323–337.

Van der Meer, T., & Tolsma, J. (2014). Ethnic diversity and its effects on social cohesion. *Annual Review of Sociology, 40,* 459–478.

Weber, M. (1922). *Economy and society: An outline of interpretative sociology.* Berkeley: University of California Press.

Wessendorf, S. (2014). 'Being open, but sometimes closed'. Conviviality in a super-diverse London neighbourhood. *European Journal of Cultural Studies, 17,* 392–405.

Wilson, H. F. (2016). On geography and encounter: Bodies, borders, and difference. *Progress in Human Geography,* published online before print, doi: 10.1177/0309132516645958.

2 Mundane mutualities

Solidarity and strangership in everyday urban life

Mervyn Horgan

An opening illustration

I first became sensitized to the curious inflections, expressions, and denials of solidarity to be found in cities while conducting ethnographic work on gentrification in Parkdale, a neighbourhood in Toronto's west end. Parkdale is best known among Torontonians for its comparatively high density of rooming houses,[1] particularly in the wake of psychiatric deinstitutionalization in the 1960s and 70s (Whitzman, 2009). Over the last two decades, Parkdale has become well-known for its steady gentrification (Slater, 2004, 2005) by Toronto's burgeoning middle class (Hulchanski, 2010). The recent history of the neighbourhood is animated by various conflicts around housing issues (e.g. the presence of squatters, the 'bachelorette crisis', a dwindling stock of privately held affordable accommodation, absentee landlords and a variety of struggles over zoning).

Gentrifying neighbourhoods bring structurally differentiated populations into close physical proximity, and therefore magnify and intensify a variety of social divisions. This provides sociologists of everyday life with a plethora of data. As part of a broader ethnography of the neighbourhood I conducted a series of 15 interviews with area residents, business owners and service providers. Among these were semi-structured interviews with homeowners who had moved into the area since 2002 and with renters who had lived in rooming houses in the neighbourhood for at least three years. These interviews lasted for between one and a half and three hours, and covered a broad range of topics including rental conditions, neighbourhood associations, landlords, property markets, renovations, zoning, schools, parks, restaurants, social services and interactions with other residents. Going into these interviews I was interested – naively perhaps – in finding evidence of positive cross-class interaction (Caulfield, 1989), maybe even friendship, between precariously housed renters in rooming houses and new middle-class homeowners buying and deconverting those same rooming houses into single family homes.

Over the course of the interviews with homeowners, something curious emerged as I began to discern a very particular kind of invocation of the term 'neighbour'. It became clear that homeowners' use of the term was highly selective, and did the work of boundary production and maintenance by explicitly expressing solidarity with some while denying it to others. This kind of selective identification is perhaps characteristic of neighbourhoods where new kinds of social mix and superdiversity emerge (see de Wilde, this volume). In the interview quotations that follow I want to draw special attention to the ways that the term 'neighbour' figures here as this bears on the conceptual elaboration that follows.

> Lisa (late 30s, homeowner since 2004): You have to be really open minded to live in this type of neighbourhood, so all of the people who live in your neighbourhood are of the same nature.
> Dave (late 30s, homeowner since 2004): Our neighbours are great – you know they are the same kind of people as you.
> Sally (early 40s, homeowner since 2003): It's kind of a weird mix for us, 'cause had we not had such great neighbours I'm not sure if we could have handled living here.
> Maggie (late 30s, homeowner since 2001): If you're buying here the chances are that your neighbours are exactly like you.

Absent from these discussions of neighbours are a large proportion of their actual physical neighbours, namely, rooming house residents (colloquially known as 'roomers'). Of the four interviewees quoted above, three live on the same block as a rooming house, and two of these live within three properties of a rooming house. An implicit distinction between neighbour and roomer emerges only in the act of articulating of who is close and who is not, who is *recognized* as a neighbour, and who is not. Across all interviews, when gentrifiers talk about their neighbours, they use the term almost exclusively for other residents who own property. This is the case regardless of duration of residency, ethnicity, sexuality or marital status. As the interview quotations suggest, spatial proximity alone does not mean that residents become neighbours, and this indicates to us something about the contours of solidaristic ties in this ethnically diverse, mixed-income neighbourhood. New homeowners in Parkdale have been steadily displacing rooming house residents over the last two decades, but those that I interviewed either ignored this pattern of displacement or only made reference to it in the abstract. For gentrifiers, neighbours form an elective community of like-minded people with similar structural positioning living in spatial proximity who share some sort of friendly relations with one another. Some who live close by are neighbours, while others are not.

The relative absence of roomers from the narratives of new property owners exemplifies Simmel's (1971) deceptively simple observation that relations between strangers are characterized by physical proximity and social distance. While rooming house residents may be neighbours in a purely physical sense (in that they live in physically proximate properties), in a social sense, roomers are distant, even negated. The gentrifiers that I spoke to were largely indifferent to the plight of roomers. Conversely, rooming house residents were well aware of the patterns of purchase and displacement that came to characterize the neighbourhood through the first decade of the 2000s. Robert, a long-term neighbourhood resident who moved regularly between rooming houses, anticipated the creeping absence – both physical and discursive – of rooming house residents from the neighbourhood: 'Roomers,' he said, 'are becoming rumours.'

Though I will not discuss further specifics of this neighbourhood here, I ask the reader the keep the above in mind as it provides a jumping off point for addressing the central theoretical concerns of this chapter. With this initial illustration my intention is to sensitize the reader to the ways that local level transformations bring structurally differentiated populations into close spatial proximity. But rather than decry the failure of some neighbourhood dwellers to generate explicit forms of solidarity, I take the utterances of gentrifiers to dramatize the particular kinds of orientations characteristic of the close physical proximity of sharp structural differences to be found in cities more generally.

Towards a minimalist conception of solidarity

Binding solidarity to particular spaces poses a quandary for the necessary and desirable expansion of solidarity more broadly. Connections between the city's spatial order and social order must be reimagined if we are to untether a quest for socio-spatial justice from a regressive territorialism. In classical sociological analysis, cities are said to have shed the *Gemeinschaftlich* ties that bound each to the other by territory, kin, shared heritage and a simple divison of labour (Durkheim, 1964; Tönnies, 1963). In this view, solidarity in cities becomes abstract, diffuse, tenuous, easily breached, but also somehow durable. Indeed, urban life exemplifies the fact that persons do not need to be friends or even personally acquainted to feel some sort of solidarity.

Under conditions where persons are strangers to one another, but are in one another's immediate coprcscnce – what I call spaces of strangership, such as urban public spaces (see Horgan, 2012; Horgan & Kern, 2014) – solidarity is enacted or its absence is demonstrated through the dynamics of interpersonal interaction. Here, I scrutinize the very basic elements of interaction between strangers in cities, to demonstrate how these interactions arc for the most part both based upon and secure some modicum of solidarity. The very idea of social order in urban space depends upon the capacities of persons who are strangers to one another to dwell together. In this chapter I treat the interpersonal negotiation of this proximity as a general concern in urban life, with the overarching aim of developing a *minimalist conception of solidarity*. With this minimalist conception I seek to outline the very barest conditions necessary for solidarity to be present in interactions between people who are strangers to one another in everyday urban life. What I offer then is an understanding of the basic features of everyday interaction that make this dwelling together possible and perhaps even desirable.

To develop this minimalist conception of solidarity, I want to focus analytic attention on the dynamics of interpersonal interaction between strangers in cities. This kind of implicitly expressed minimal form of solidarity (what we might call soft solidarity) can be distinguished from those forms of solidarity institutionalized by states through citizenship rights and welfare entitlements, and the formally organized kinds of solidarity harnessed and cultivated by civil society groups, social movements and labour unions. The kind of solidarity of interest to us here is neither explicitly expressed nor codified in law, but rather is implicitly conveyed through the intersubjective accomplishment of ordinary everyday urban life.

This turn to understanding the everyday enactment and achievement of solidarity redirects attention away from the formal institutionalization of solidarity in societies writ large, aiming instead to illuminate the intersubjectively elaborated interactional norms and conventions that organize orientations and interactions between strangers in such a way as to *informally institutionalize* solidarity in everyday urban life. Developing this minimalist conception of solidarity, then, offers a way to theorize the conditions under which solidarity between strangers in everyday urban life might be made and unmade. In addition to this specific conceptual contribution, and in line with the strand of general social and political theory that treats solidarity as a precondition for justice (Alexander, 2006; Kymlicka, 1995), developing a minimalist conception of solidarity is firmly bound up with a broader concern with understanding the necessary conditions for the achievement of social justice. By exploring the most basic foundations of solidarity in urban sociality, I offer a conceptual elaboration of some of those enduring aspects of our collective predicament that undergird any possibility of advancing solidarity in a society of strangers.

To do this, I use elements of everyday urban life to bridge two very different literatures tangentially concerned with solidarity: (i) critical theoretical work on recognition; and (ii) interpretive sociological work on urban interaction. This will create some conceptual space to sketch an outline of the minimal conditions necessary for the generation and sustenance of solidarity among strangers in cities. With these insights from critical and interpretive social theory in hand, I demonstrate how everyday solidarity between strangers in cities is created, sustained and dissolved. Rather than asking how we might *produce* solidarity between strangers (a question that assumes no pre-existing solidarity), I ask instead how might we conceive of an *already existing* – albeit weak – form of solidarity. The kind of solidarity of concern here is spatially circumscribed in that it requires copresence, but is not bound to territory in the way that, for example, nation-based solidarities are. Of interest here is the kind of solidarity that is key to the mundane production of social order at the level of everyday life in the city.

Developing a sociologically grounded conception of the interactional conditions that make the most minimal kinds of recognition and solidarity both possible and necessary requires some conceptual work. Below I advance the concept of the *urban interaction order* by formulating strangership as a core element of urban sociation and social order. Here I show how the *mutual indifference* that broadly characterizes interactions between strangers in cities constitutes and sustains this order, and counterintuitively, how this apparently non-interactional form of interaction is threaded through with solidarity. Cities provide the socio-spatial context for the achievement of strangers' mutual indifference to one another, and I show how the solidarity embedded in mutual indifference makes dwelling together in cities possible. By way of conclusion, I argue that the consistently patterned demonstrations of solidarity that are largely taken-for-granted in everyday urban life depend upon and instantiate what I call *minimal mutual recognition*. Though minimal, this kind of recognition is a baseline for the achievement of social justice, and so where these minimal conditions are not met, solidarity is undermined and injustice is likely to prevail.

Recognizing strangers

Scholarship around the concept of recognition has burgeoned across the humanities and social sciences over the last quarter century. Philosophically inspired by a reinvigoration of Hegelian ethics, a range of social and political theorists and philosophers have sought to foreground the ways that taking the principle of recognition seriously can bring us to think anew about justice in the contemporary epoch (Honneth, 1996, 2007; Fraser, 2001, 2009; Fraser & Honneth, 2003; Petherbridge, 2011, 2013; Taylor, 1994). That said, despite the fact that the central and explicit concerns of recognition scholars revolve around the big questions of solidarity, justice, inclusion and diversity, contemporary recognition scholarship tends to operate either as a highly abstracted conceptual system or within a juridicio-legal framework focused on elucidating the formal-legal mechanisms that might bolster recognition across the political space of the nation or through transnational institutions (Benhabib, 2002; Fraser, 1997, 2009; Kymlicka, 1995; Taylor, 1994). Moreover, very little of this work has had an urban focus, even though it could provide a useful counterweight to urban political economy's focus on redistribution as the primary mechanism for achieving social justice.

Where recognition scholarship tends towards the abstract, the normative and the formal, microsociology orients itself to the empirically grounded, phenomenological description of mundane elements of collective life (Goffman, 1963, 1983; Rawls, 1987, 2009; Schutz, 1967), focusing in particular on the minutiae of everyday encounters. The richly textured understanding of the intersubjective dimensions of social life offered by microsociologists generally comes at the cost of sidestepping the larger macrostructural contexts that both condition and organize human copresence. There are some possibilities for rapprochement between these two literatures, and in this spirit I offer some conceptual tools for deepening and extending our understanding of solidarity as it is achieved in face-to-face interaction between strangers at the level of the city street in the here and now.

In bridging recognition scholarship and microsociology, we traverse fertile but underexplored conceptual terrain for theorizing the ordinary everyday production, maintenance and dissolution of solidarity among strangers in cities. It is in foregrounding some of the normative underpinnings of microsociological work that we can provide stronger sociological foundations for the normative claims of recognition scholarship, which I read here as implicitly concerned with the expansion of solidarity. Working to achieve this goal through a careful analysis of the mundane realities of everyday urban life as one lived among strangers will go some way towards addressing present-day scepticism around the possibilities of what Ash Amin (2012, p. 80) has called an 'urban politics of recognition'.

For our purposes, contemporary recognition scholarship is inadequate. On the one hand, a formal institutionalist bias mirrors in part a similar bias in contemporary work on solidarity (Pensky, 2008). On the other hand, there is a strand of recognition scholarship that does focus on intersubjectivity, but this is largely centred on intimate relations (see Honneth, 1996, pp. 92–130). This has brought Lois McNay (2008) to point out the shortcomings of recognition theory vis-à-vis 'social weightlessness', in reference to this literature's failure to attend adequately to the lived experience and reality of recognition and its absence. Undoubtedly, a fundamental condition for self-worth, self-esteem and the development of the self more generally derives from intersubjective recognition in our most intimate relations. Likewise, recognition of individuals by external institutions is essential in providing security of the person as is the case in the citizen–state relation. Nonetheless, a great range of social life occurs between the poles of formal institutions oriented to citizens and interest groups in general, and intimate relations oriented to the absolute specificity of those we love. In general, recognition theory does not provide for the possibility of the intersubjective achievement of recognition between copresent *strangers*, that I argue can be discerned in everyday urban life.

By restricting ourselves to currently prevailing conceptions of recognition, we risk prematurely foreclosing opportunities for the development of solidarity at other levels and through alternative affective registers, for example, not only between intimates, but also between consociates more generally. If recognition remains a philosophical bone awaiting sociological flesh, then working with the concept of solidarity can furnish recognition with some 'social weight'. This is where we can begin to draw on some microsociological ideas.

Spaces of strangership

'Relation is mutual.' (Buber, 1958, p. 15)

Just as a friendship refers not to an individual, but rather to the relationship between friends, so too strangership refers to the relationship between strangers (Horgan, 2012). In treating cities as spaces of strangership, I focus attention on *relations* between strangers in cities, rather than on the specific characteristics of any particular individual or group. With this in mind, we can now redirect attention away from the formal institutionalization of solidarity and work through a cluster of interrelated concepts – anonymity, mutual indifference and the urban interaction order – that together depend on a minimal form of interpersonal recognition which, I argue, is integral to the production and maintenance of solidarity under conditions of strangership. These concepts are clearly interconnected, so I distinguish them here primarily for heuristic purposes to open up what I think are some useful ways to think anew about cities as spaces of strangership. First, I will deal briefly with anonymity and the urban interaction order, before moving to a more detailed discussion of mutual indifference as a solidarizing interactional mechanism, albeit a counterintuitive one.

As the condition or state of being unknown to or by others, anonymity is clearly a central element of urban life, one that opens up possibilities of personal freedom and the expansion of individuality, while simultaneously raising the peril of social isolation and the fragmentation of community (Klinenberg, 2001; Putnam, 2001). The most ubiquitous form of association we find in cities is between strangers. Consequently, anonymity is the most prevalent individual condition in everyday urban life. So, even while the anonymity and isolation of the individual may be intuitively connected, any individual's anonymity is *socially* created and sustained.

One remarkable feature of everyday urban life is that despite the fact that we move among others who are unknown to us, for the large part, some form of order is maintained. I refer to the order produced by and through the everyday copresence of strangers in cities as the *urban interaction order*. I call this an order for the simple reasons that it is organized in ways that make ordinary urban life possible, and because it is structured by specifiable principles and regularities of functioning. I treat the urban interaction order as a situationally emergent but stable and enduring feature of everyday social interactions between strangers. In developing this concept, I draw on the work of dramaturgical sociologist, Erving Goffman who coined the term 'interaction order'.

For Goffman (1983, p. 2), the interaction order comprises the domain of face-to-face interaction where persons are physically copresent, noting that it can be analysed 'as a substantive domain in its own right'. The interaction order, then, is the *situated* social space within which we *encounter* one another in every part of our lives where we are *copresent* with others. Important to note here is that the interaction order does not depend on explicit interaction between persons: the mere fact of human copresence has the remarkably enduring quality of producing some form of order in more or less all but the most extreme cases.

Like Goffman, I afford primacy to relations and interactions *between* copresent individuals rather than the supposedly inherent characteristics of any one individual. By adding the modifier *urban*, I delineate a specific kind of spatial and experiential domain that draws from those who populate and sustain it a range of personal adjustments and orientations not necessarily generalizable to the more all-encompassing interaction order. The *urban* interaction order, then, encompasses the assortment of unceasing yet very ordinary and taken-for-granted encounters between anonymous individuals who are strangers to one another in cities.

While there is much that we could do analytically with the urban interaction order as a substantive domain, my primary interest here is in understanding the core interactional mechanisms and processes that underpin this order, for it is here that we can begin to trace out the lineaments of our minimalist conception of solidarity. All interaction orders depend on particular kinds of orientations from those who populate and enliven them, such that particular kinds of orientations become characteristic of particular kinds of interaction order. Think, for example, of the different kinds of orders and orientations characteristic of a classroom, compared to a dinner party, compared to a city street at rush hour.

The urban interaction order requires a specific form of reciprocal orientation, what classical social theorists Émile Durkheim (1964, p. 298) and Georg Simmel (1971, p. 334) both call *mutual indifference*. While the apparent indifference of urbanites to one another typically grounds popular criticisms of urban living, given our interest in solidarity what is important to remember here is that this indifference is *mutual*. So, while the two elements of this term may appear at first glance to be somewhat contradictory, they modify one another in significant ways. Indifference, means to be uninterested in, unconcerned with, or unmoved by the presence of others, while mutuality implies a shared orientation or a kind of reciprocity. These two different – but not opposed – elements combine, so that mutual indifference characterizes social relationships in which both parties recognize one another but only in the most diffuse and unconcerned way. Mutual indifference is the core characteristic of social relationships between anonymous individuals that sustains the urban interaction order. Further, mutual indifference is the central relational mechanism that maintains *both* anonymity for individuals and the urban interaction order for the collective.

In sum then, anonymity is individually experienced but socially and situationally sustained. The urban interaction order is supraindividual but dependent on individuals in general for its sustenance. And mutual indifference is the intersubjectively elaborated, dominant form of interaction between copresent strangers.

Maintaining mundane mutualities

In mutual indifference we find a social relationship, albeit a somewhat confounding one (Weber, 1978, pp. 27–28). Strangers in cities are complicit in agreeing to only interact in aloof, disattentive ways (Lofland, 1973, p. 155). To become an urban dweller, one must learn to negotiate the unspoken agreement that, for the most part, spatial proximity will not lead to overt social contact. This is what Goffman (1963, pp. 83–88) refers to as 'civil inattention'. Relations of mutual indifference, then, *require* individuals to be civilly inattentive.

By upholding the implicit pact between individuals to not engage one another *as* individuals, mutual indifference sustains a social relationship that demonstrates to us that, even in the midst of anonymous urban spaces, some modicum of social solidarity can be discerned. I do not wish to claim that this kind of solidarity is a model for successful societies or political citizenship, but rather, that the collective maintenance of the urban interaction order suggests to us that even in contexts where explicit connections between persons are absent, social solidarity can be found, albeit in a weak form. Solidarity is found here in the complicity of urban dwellers in working to make it apparent that they are indifferent to one another. When urban dwellers provide one another with the 'minimal courtesy' (Goffman, 1963, p. 84) of civil inattention, solidarity is jointly offered and honoured through mutual indifference.

As the relational mechanism that binds individuals to one another within the urban interaction order, mutual indifference operates at both an individual and a collective level. For the individual, indifference is borne in large part out of the dominance of instrumental rationality as a response to the experiential and interactional density of city space; it is in this sense functional, since the intensity and variety of urban experience – what Simmel (1950, p. 410) famously called 'the intensification of nervous stimulation' – requires specific individual adaptations to the urban environment (see also Durkheim, 1964, p. 298; Tonkiss, 2003). Thus, indifference is part of the repertoire of individual dispositions best suited to the intensity of urban life such that relations between anonymous individuals *must* be mostly characterized by indifference.

At the level of the collective, mutual indifference is foundational to the urban interaction order. Curiously then, indifference – at least when mutual – is less a symptom of decline, disintegration, malice and disorder, than it is a widely and implicitly supported interactional mechanism that permits us to live and persist among strangers. It is the generalized distribution and diffusion of this mechanism across the social space of the city that gives social solidarity in the city its peculiar character. Interpersonal encounters between strangers within the urban interaction order are organized by an implicitly shared commitment to the maintenance of that order. As an interactional mechanism, mutual indifference is essential to the generic social process by which the urban interaction order is collectively produced. This means that the surface appearance of indifference as inaction or some pathological form of urban individualism can be sociologically reformulated as an expression of a shared commitment to the collective production and protection of the urban interaction order.

While mutual indifference is the basic interactional mechanism upholding the urban interaction order, treating it in a purely conceptual way risks veiling the messiness and contingency of everyday life, what Goffman calls its 'promissory character' (Goffman, 1983, p. 2). So, while mutuality presupposes a principle of symmetry – if not structural, at least interactional – not all urban social relations can be said to be organized around this principle. Therefore, it is worth considering a form of indifference that is *non-mutual*, and in particular, attending to how this bears on solidarity more generally.

Mutuality undone? Non-mutual indifference and inequality

As elaborated above, mutual indifference is held together at its core by a fundamental equality between city dwellers. Those who are mutually indifferent to one another are complicit: their relationships, if not structurally symmetrical, are at least interactionally so. But this is the case only so long as their indifference is mutual. While mutual indifference may characterize mundane everyday urban life, as noted at the outset, in cities sharp structural differences frequently dwell cheek by jowl, bringing inequality and injustice into sharp relief. The maintenance of the urban interaction order, then, requires some bracketing of the range of inequalities that the city brings to light. Indeed, the intensification of inequality in cities around the world may work to undermine the kind of interactional mutuality that I posit as central to producing and sustaining a minimal form of solidarity under conditions of strangership. Though mutual indifference is central to the urban interaction order, in urban space we also find non-mutual indifference. Existing structural inequalities can be enunciated through this latter form of indifference.

Back then to the illustration from the gentrifying Toronto neighbourhood that opened this chapter. Returning to the interviews quoted above, one can recall that middle-class homeowners expressed some affection for and affinity with their neighbours, but clearly used this term – neighbours – to refer to others like them and not to rooming-house residents. Consider the kind of interactional life commensurate with the disappearance of rooming-house residents from the narratives of new property owners. The sort of narrative erasure found in gentrifiers' descriptions of their neighbourhood is, I think, both an instance of and symptomatic of a confluence between structural inequality and non-mutual indifference.

With this in mind, we might then posit that urban social relations between strangers operate on a continuum between mutual and non-mutual indifference, where the degree of mutuality corresponds to the extent to which each individual or group both benefits from and is complicit in sustaining indifference. Indifference is mutual if it permits each to go about their everyday lives unencumbered by others; it is non-mutual, or asymmetrical, if one party is negatively impacted in the course of another's indifference. Structural forces exogenous to an interaction do, of course, influence its course and its contents, but these forces alone cannot determine whether or not any specific relation between strangers is to be characterized by mutual or non-mutual indifference. Following Goffman, we can treat the urban interaction order as having some autonomy from broader social structures, as the production of local order by copresent persons cannot be wholly determined by exogenous forces.[2] Mutual indifference evidences the kind of everyday solidarity that upholds the urban interaction order, but indifference, when characterized by asymmetry – non-mutuality – (re)produces the city as an exclusionary social form. Here, solidarity is undone.

Non-mutual indifference reproduces and sustains existing structural inequalities, and produces new interactional ones. It depends on fundamentally asymmetrical orientations that undermine and even sever those bonds, however weak or abstract, upon which urban social solidarity depends. Solidarity is absent where an asymmetry of structural positions overrides and/or militates against the mutuality of orientations that necessarily characterize the urban interaction order. When indifference ceases to be mutual – when it is non-mutual or asymmetrical – then the principle of mutual recognition key to urban social solidarity in particular, and to social solidarity more broadly, is lost.

The mutual indifference upon which the urban interaction order rests is generally unproblematic, in the sense that the goals – if we can call them that – of a mutually indifferent relation are reached when problems do not arise, when nothing intervenes to alter that relation, when this mutual indifference benefits, is desired by, or is necessary to *both* parties. Embedded in mutual indifference, then, is a kind of mutual recognition, albeit low-intensity and weak or, what we are calling minimal. If indifference is mutual – that is, if it is based on minimal mutual recognition – then the generic social process and interactional mechanisms that instantiate solidarity and uphold the urban interaction order are in operation. Conversely, if indifference is non-mutual, one individual or group is disproportionately impacted by the indifference of the other individual or group.

In this sense, non-mutual indifference produces situational inequality that not merely reflects, but propagates and reinforces structural inequality. In the example of the gentrifying neighbourhood, while there are obvious differences in the formal-legal rights that accrue to owners versus renters, there also appears to be a more subtle, but nonetheless powerful, kind of inequality that plays out in the relations between them. Hence, non-mutual indifference can operate analytically as a sensitizing concept that provides access to processes of solidarization and desolidarization in everyday urban life.

Conclusion

This chapter connected the most mundane practices of everyday life to the abstract philosophical formulations of recognition as a key dimension of social justice in particular, and of collective life more generally. To do this, I outlined how two broad literatures – recognition scholarship and microsociology – might complement one another in order to advance our understanding of solidarity between strangers in everyday urban life. Above, I have shown how, when brought together, work in these areas can be of mutual benefit. More specifically, I analysed the functioning of mutual indifference in everyday urban life to show how it forms a key element of social solidarity between strangers.

As mentioned earlier, following Alexander (2006, p. 13; see also Kymlicka, 1995, pp. 173–174), I take it that '[j]ustice depends on solidarity', and so, the kind of solidarity that I discern in everyday urban life can, I argue, also be read as a basic precondition for justice. The uneven and asymmetrical distribution and enactment of indifference discussed in this chapter foregrounds questions of solidarity and so also of justice. The production of bounded solidarities necessarily implies a limit beyond which justice is not extended: justice is limited by the boundaries of solidarity. My counterintuitive claim here is that mutual indifference – as a kind of social distanciation that depends on the capacity to give another 'just-enough' recognition, a minimal engagement but without deeper interference – is key to just forms of everyday urban life in the conditions of high population density, strangership and 'super-diversity' (Vertovec, 2007) that increasingly characterize contemporary societies. By no means do I intend to suggest that mutual-indifference-as-recognition supercedes legal equality and substantive rights, but enshrining the 'right to the city' (Harvey, 2003, 2008), for example, in law alone by no means guarantees justice and solidarity at the level of everyday life. While formal and substantive justice must be achieved, this is not possible if more mundane or quotidian forms of solidarity are not advanced too. Given the intertwining of justice and solidarity, the achievement of formal justice will be barren unless the kind of everyday solidarity analysed and advocated for here also flourishes. This flourishing rests at the very least on minimal mutual recognition.

Avenues for future research in this area are wide. Examining the extent to which the distribution of non-mutual indifference across the social space of the city coincides with discriminatory and/or exploitative practices would likely be fruitful. More generally, investigating where non-mutual indifference is prevalent, and among whom, is likely to yield insights into the everyday production and experience of inequality. These are rich veins for further inquiry that will undoubtedly help us to deepen and extend our understanding of the dissolution of solidarity in everyday life and, by extension, assist us too in developing strategies to address the persistence and exacerbation of injustice in cities.

That cities are characterized by indifference is the source of much of the disdain that people express towards urban life. As I have demonstrated here, indifference is a form of social action which must be more fully explored so that we might develop a deeper analytic understanding of the way it shapes social relationships and, counterintuitively, how indifference, when it is mutual, plays a key role in sustaining solidarity between anonymous strangers in everyday urban life.

Notes

1 In Toronto, 'rooming house' is the offical designation for what is called single room occupancy (SRO) housing or bedsit in other jurisdictions. Each unit in a designated rooming house can legally house a single person with a small kitchenette and shared washroom facilities.
2 This is not to downplay the power of social structure in everyday life, but rather to posit that the endogenous organization of the urban interaction order can be *analysed* independently of social structure.

References

Alexander, J. (2006). *The civil sphere*. New York: Oxford University Press.

Amin, A. (2012). *Land of strangers*. Cambridge: Polity.

Benhabib, S. (2002). *The claims of culture: Equality and diversity in the global era.* Princeton, NJ: Princeton University Press.

Buber, M. (1958). *I and thou*. New York: Charles Scribner's Sons.

Caulfield, J. (1989). Gentrification and desire. *Canadian Review of Sociology and Anthropology, 26,* 617–632.

Durkheim, E. (1964). *The division of labor in society*. New York: Free Press.

Fraser, N. (1997). *Justice interruptus: Critical reflections on the 'postsocialist' condition.* New York and London: Routledge.

Fraser, N. (2001). Recognition without ethics? *Theory, Culture & Society, 18,* 21–42.

Fraser, N. (2009). *Scales of justice: Reimagining political space in a globalizing world.* New York: Columbia University Press.

Fraser, N., & Honneth, A. (2003). *Redistribution or recognition? A political-philosophical exchange*. London: Verso.

Goffman, E. (1963). *Behavior in public places: Notes on the social organization of gatherings*. New York: Free Press.

Goffman, E. (1983). The interaction order. *American Sociological Review, 48,* 1–17.

Harvey, D. (2003). The right to the city. *International Journal of Urban and Regional Research, 27,* 939–941.

Harvey, D. (2008). The right to the city. *New Left Review, 53,* 23–40.

Honneth, A. (1996). *The struggle for recognition: The moral grammar of social conflicts.* Cambridge, MA: MIT Press.

Honneth, A. (2007). *Disrespect: The normative foundations of critical theory*. Cambridge: Polity Press.

Horgan, M. (2012). Strangers and strangership. *Journal of Intercultural Studies, 33,* 607–622.

Horgan, M., & Kern, L. (2014). Urban public spaces: Streets, strangership and securitization. In H. Hiller (Ed.), *Urban Canada* (pp. 112–132). Toronto: Oxford University Press.

Hulchanski, J. D. (2010). *The three cities within Toronto: Income polarization among Toronto's neighbourhoods, 1970–2005*. Toronto: Cities Centre, University of Toronto.

Klinenberg, E. (2001). Dying alone: The social production of urban isolation. *Ethnography, 2,* 501–531.

Kymlicka, W. (1995). *Multicultural citizenship: A liberal theory of minority rights*. Oxford: Oxford University Press.

Lofland, L. (1973). *A world of strangers: Order and action in urban public space*. New York: Basic Books.

McNay, L. (2008). *Against recognition*. Cambridge: Polity.

Pensky, M. (2008). *The ends of solidarity: Discourse theory in ethics and politics*. Albany: State University of New York Press.

Petherbridge, D. (2011). *Axel Honneth: Critical essays*. Boston, MA: Brill.

Petherbridge, D. (2013). *The critical theory of Axel Honneth*. Lanham, MD: Lexington Books.

Putnam, R. D. (2001). *Bowling alone: The collapse and revival of American community*. New York: Simon & Schuster.

Rawls, A. W. (1987). The interaction order *sui generis*: Goffman's contribution to social theory. *Sociological Theory, 5,* 136–149.

Rawls, A. W. (2009). An essay on two conceptions of social order: Constitutive orders of action, objects and identities vs aggregated orders of individual action. *Journal of Classical Sociology, 9,* 500–520.

Schutz, A. (1967). *The phenomenology of the social world*. Chicago, IL: Northwestern University Press.

Simmel, G. (1950). The metropolis and mental life. In K. Wolff (Ed.), *The sociology of Georg Simmel* (pp. 409–424). New York: Free Press.

Simmel, G. (1971). *On individuality and social forms*. Chicago, IL: University of Chicago Press.

Slater, T. (2004). Municipally managed gentrification in South Parkdale, Toronto. *The Canadian Geographer, 48,* 303–325.

Slater, T. (2005). *Toronto's South Parkdale neighbourhood: A brief history of development, disinvestment, and gentrification* (Research Bulletin #28). Toronto: Centre for Urban and Community Studies.

Taylor, C. (1994). *Multiculturalism: Examining the politics of recognition*. Princeton, NJ: Princeton University Press.

Tonkiss, F. (2003). The ethics of indifference: Community and solitude in the city. *International Journal of Cultural Studies, 6,* 297–311.

Tönnies, F. (1963). *Community and society*. New York: Harper Torchbooks.

Vertovec, S. (2007). Super-diversity and its implications. *Ethnic and Racial Studies, 30,* 1024–1054.

Weber, M. (1978). *Economy and society: An outline of interpretive sociology*. Berkeley: University of California Press.

Whitzman, C. (2009). *Suburb, slum, urban village transformations in Toronto's Parkdale neighbourhood, 1875–2002*. Vancouver: UBC Press.

3 Learning to cope with superdiversity

Place-based solidarities at a (pre-)primary Catholic school in Leuven, Belgium

Nick Schuermans and Pascal Debruyne

Introduction: the school as a place of encounter and solidarity

In the last two decades, social geographers, sociologists, anthropologists, linguists and educationalists have embarked upon a joint project to disentangle the lived realities and the complexities of social relations in superdiverse places (e.g. Amin, 2002, 2012; Vertovec, 2007; Wise & Velayutham, 2009; Arnaut, 2013; Nowicka & Vertovec, 2014). Much research in this field is conducted around the basic assumption that occasional exchanges in public spaces and informal interactions at work, at school or at leisure allow people with different migration histories, cultural backgrounds and/or socio-economic situations to learn to live with each other's differences (e.g. Binnie et al., 2006; Valentine, 2008; Wessendorf, 2013; Schuermans, 2016; Wilson, 2016).

Among sociologists of education, much attention has been paid to the potential effects of superdiversity at school (e.g. Moody, 2001; Quillian & Campbell, 2003; Mouw & Entwisle, 2006). Often inspired by Allport's (1954) contact hypothesis or Blau's (1977) structural theories of social interaction, they are interested in the relationships between the ethnic composition of schools, interethnic contacts, interethnic friendships and ethnic prejudices. Despite fundamental differences in the operationalization of dependent and independent variables, large-scale quantitative studies generally imply that attitudes towards other ethnic groups are more positive in mixed schools because there are more opportunities for meaningful interactions across ethnic lines (Hallinan & Williams, 1989; Thijs & Verkuyten, 2014). Based on survey research in 85 secondary schools in Flanders, Van Houtte and Stevens (2009) calculated, for instance, that Flemish students in schools with a greater proportion of immigrant students have more friends with an immigrant background. In their study on late adolescents, Dejaeghere et al. (2012) warned, however, that interactions between different ethnic groups only bring about significantly lower levels of prejudice if there is a positive climate between the different groups.

Stimulated by a growing interest in the geographies of education (Collins & Coleman, 2008; Holloway et al., 2010; Cook & Hemming, 2011), human geographers are increasingly paying attention to everyday interactions at diverse spaces of education as well. Mainly drawing on in-depth interviews and ethnographic fieldwork with children and their parents, they reveal that multicultural social networks develop in multicultural schools and that an awareness of cultural differences, religious sensitivities and language issues can inform more socially inclusive practices (Kong, 2013; Neal & Vincent, 2013). By way of example, Noble's (2013) ethnographical study in a handful of multicultural schools in middle-class suburbs of Sydney points at the habits and routines that fashion intercultural dialogues and cosmopolitan dispositions at school. Hemming's (2011) extensive fieldwork in English primary schools confirms that diverse classrooms and playgrounds allow children from different cultures and religions to understand each other's habits and to develop friendships outside their own communities. Ethnographic research in a multicultural primary school in Birmingham taught Wilson (2014) that many encounters with Muslim mothers and children are characterized by discourses of loss – or even betrayal – of British norms and values, but that new lines of commonality and solidarity can develop around shared parental commitments as well.

While both geographers and sociologists have mainly focused on the encounters of children and parents at multicultural schools, surprisingly little attention has been paid to the experiences and the practices of the professionals who give shape to the everyday life at school. Taking into account that very few teachers and school principals actually belong to ethnic minorities in many countries across the world, it needs to be questioned how these professionals experience their daily encounters with the diverse school population and how these encounters help them to reconsider long-standing ways of talking about – and dealing with – pupils and parents who are not like them. In our view, this is a fundamental question. A lot of research indicates that the attitudes of teachers and school leaders, student–teacher relationships and inclusive school identities have an important influence on the interethnic relations between students at school (Van Avermaet & Sierens, 2010; Agirdag et al., 2013; Thijs & Verkuyten, 2014; Santagati, 2015). Many scholars have also reached the pessimistic conclusion that institutional racism affects the well-being, the school results and the social mobility of many school-going children (e.g. Brooks, 2012; Araújo, 2016).

By focusing on the nature and the effects of the encounters of teachers, school principals and other professionals working at a Catholic primary and pre-primary school in Leuven (Belgium), this chapter aims to contribute to sociological and geographical studies on interethnic contacts at school – and beyond – in at least three ways. First, much of the literature understands the potential effects of interethnic encounters as an increased respect for difference (e.g. Hemming, 2011), lower levels of ethnocentrism (e.g. Dejaeghere et al., 2012) or the disruption of existing knowledges and ways of living (e.g. Wilson, 2014). In line with a variety of approaches that relocate social agency in actions and performances rather than in discourses and debates, it is essential to complement research on how encounters change the way people think and talk with studies on what people actually do. After all, there is a serious risk that the values, the attitudes and the stereotypes people express in interviews do not match the actual practices they deploy in their everyday lives (Valentine, 2008, p. 323). To overcome the potential mismatch between values and behaviour, we opt to study the practices teachers and other professionals unfold at school, rather than their opinions, attitudes and discourses. Defining solidarity as 'the willingness to share and redistribute material and immaterial resources drawing on feelings of shared fate and group loyalty' (Stjernø, 2004, p. 25), we are particularly interested in practices of solidarity deployed by professionals at school.

A second contribution to the wider literature on contact across ethnic and socio-economic lines relates to the geography of the effects of encounters. Through the adoption of a relational understanding of place, we aim to demonstrate that solidarities which are nurtured through encounters in a particular place do not necessarily have to be bounded to that place (see Massey, 2004, 2005). Rather than limiting our focus to the micro-geographies of contact and solidarity within the school, we demonstrate how teachers and other school staff build upon their personal networks and professional expertise to mobilize solidarities at various spatial scales. By situating the case study of the school – and the solidarities that develop there – within the larger scales of the neighbourhood, the city and the Flemish region, we aim to demonstrate how encounters can have an impact beyond the spatial and the temporal immediacy of the here and now (see Valentine, 2008, p. 325).

A final contribution to the geographical and sociological literature on encounters concerns the causal relationship between the encounter, on the one hand, and its effects, on the other (see also Wilson, in this volume). Paraphrasing Valentine (2008, p. 325), much of the literature on encounter seems to reproduce the naive assumption that contact across lines of race, class and ethnicity automatically translates into respect for difference. Even though it is generally acknowledged that encounters are mediated by individual biographies, civic principles of interaction and very unequal power relations, few scholars have tried to unwrap the black box of this mediation effect (see Wilson, 2016). Inspired by the fourfold division of sources of solidarity elaborated in the introductory chapter of this volume, we aim to disentangle whether encounters at school are mediated by – or act as catalysts for – joint struggles for good education, relationships of interdependency and shared norms and values that promote solidarity (see Oosterlynck et al., 2016, p. 669). In this way, we intend to open up part of the black box between the moment of the encounter and its potential effect under the form of solidarity.

To make these three points, the remainder of this chapter is organized in five more sections and a conclusion. In the next section, we introduce the Flemish educational landscape, the specific school in which we conducted our study and our research methodologies. In the subsequent section, we describe the difficulties of the first encounters with a superdiverse school population. The two following sections focus on the kind of solidarities developed by different professionals working at school. First, we look at acts of solidarity at the level of the school as an institution. Afterwards, we discuss (the lack of) solidarity at larger spatial scales, ranging from the neighbourhood and the town to the Flemish region. The last empirical section discusses the sources of these solidarities. In the conclusion, we draw on these four empirical sections to come back to the three contributions to the sociological and geographical work on interethnic contacts elaborated above.

Introducing Mater Dei

Over the last decades, Flanders has rapidly become ethnically and culturally diverse. In 2015, the research department of the Flemish government calculated that foreign nationals made up 7.8 per cent of the inhabitants of Flanders (Van den Broucke et al., 2015, p. 80). In 2013, 18.4 per cent of the population was of foreign descent (Van den Broucke et al., 2015, p. 99). Whereas most immigrants in the 1960s and 1970s originated in a limited number of Mediterranean countries such as Morocco, Turkey, Italy, Spain and Greece, the current population statistics are marked by a fast diversification of the countries of origin. Especially in the larger towns and cities, the emergence of a truly superdiverse population can be observed (Maly et al., 2014). Within a decade, ethnic minorities are expected to make up the majority of the population in the city of Antwerp (Geldof, 2013, p. 27). Even smaller towns such as Leuven have inhabitants from more than 150 different countries (Stad Leuven, 2015).

Because younger age cohorts are characterized much more by ethnic, cultural and linguistic diversity than older ones, the challenges related to the development of a superdiverse society are probably felt most in the field of education. In international comparisons, such as the Program for International Student Assessment (PISA), Flemish education is generally considered to be one of the top performers, but it is also marked by one of the highest performance gaps between children of different economic, social and cultural status (Danhier et al., 2014). The underachievement of students with an immigrant background is expressed in higher dropout rates, higher levels of grade retention, their underrepresentation in higher education and their overrepresentation in less esteemed technical and vocational tracks of secondary school (D'hondt et al., 2015, pp. 336–337). All in all, the equity of the Flemish educational system ranks as one of the weakest of all developed and democratic countries (Danhier et al., 2014, p. 7). This implies that Flemish schools evidence less social mobility for students from immigrant families and poor households than their counterparts in comparable countries.

To explain the performance gap between children of different backgrounds, scholars point at different causal mechanisms, from social prejudices and discriminatory practices among teachers to language barriers and the content of textbooks (e.g. Schuermans, 2014; D'hondt et al., 2015; Pulinx et al., 2015). Many also argue that Flanders has exceptionally high levels of ethnic and socio-economic school segregation (Spruyt, 2008) and that much of the performance gap is due to the underachievement of schools with a high share of pupils from ethnic minorities and working-class backgrounds (Jacobs et al., 2009; Danhier et al., 2014). Agirdag et al. (2013) explain that the underachievement in these schools is, at least partly, due to the lower expectations of teachers. Especially when schools adopt a stringent monolingual Dutch-language policy, teachers tend to have less trust in pupils with another mother tongue and lower expectations with regards to their academic achievements (Pulinx et al., 2015). In any case, it is clear that teachers, school principals and other school staff have a big influence on the school achievements and life chances of underprivileged pupils in so-called 'concentration schools' (see Peleman, 1998).

Against this background, this chapter examines the way in which teachers, principals and other members of the school staff of a Flemish school have shown solidarity by transforming their school in response to the challenges mentioned above. Located in the university town of Leuven, about 25 kilometres from Brussels, Mater Dei is a primary and pre-primary school certified by the Catholic church. The school was established more than 150 years ago by sisters of a nearby convent as a boarding school for young novices, orphans and girls from the neighbourhood (Uytterhoeven & Van den Heuvel, 1994). While the percentage of regular churchgoers in Flanders has decreased from more than 50 per cent in the 1960s to 5 per cent in 2009 (Havermans & Hooghe, 2011, p. 29), Catholic schools still enrol more than half of all primary and almost three-quarters of all secondary school students. Just like most other Catholic schools, Mater Dei is not-for-profit and funded by the state.

Nowadays, Mater Dei serves more than 300 children between two and a half and 12 years old. In many ways, the current school population can be called superdiverse (see Vertovec, 2007). According to the most recent language statistics, as much as 83 per cent of the pupils in Leuven's primary schools speaks Dutch at home, but this is the case for less than 45 per cent of the children at Mater Dei (Agentschap voor Onderwijsdiensten, 2015). Within the group of pupils who do not speak Dutch at home, there is also a wide variety of mother tongues, citizenship status – some very precarious – and countries of origin. In a typical class, you can find children (with parents) from countries as diverse as the Philippines, Nepal, Ghana, Morocco, Turkey, Congo, Eritrea, Pakistan, Poland, Burundi, Vietnam, India and Belgium. Statistics indicate that many of these children grow up in poverty. In Flanders, parents with an income below a certain threshold are eligible for a school grant. In Leuven, 17 per cent of the school-going children are in this situation. In Mater Dei, this amounts to 34 and 52 per cent in the pre-primary and primary school, respectively (Agentschap voor Onderwijsdiensten, 2015).

In what follows, we will discuss the multiple ways in which employees of Mater Dei have tried to address challenges related to superdiversity, multilingualism and socio-economic disadvantage. As elaborated in the introduction to this chapter, we will consider some of these practices as acts of solidarity. Throughout our analysis, we will pay special attention to the sources of these solidarities and the way in which they can extend beyond the territory of the school.

To analyse the character, the geographical extent and the sources of the place-based solidarities developed by the staff of Mater Dei, we rely on different types of data. First, we carried out a document analysis of a range of different documents such as the mission statement of the school, the inspection reports of the Flemish government, the website and the equal educational opportunities plan. Additionally, we conducted 12 in-depth interviews with former and current school principals, the chairperson of the board of directors, an employee of the student guidance centre, the Leuven alderman of education and the coordinator and employee of a community organization involved in the school. In the staff-room, we also had informal conversations with members of the teaching staff. During three days of participatory observation in the classes, the lunch room and the playground of the school, we examined interactions between children and teachers. Ongoing action research focuses on the image of the school both inside and outside the school walls.

Difficult encounters with superdiversity

For many interviewees, the transition to a superdiverse school population had come as a surprise. Towards the end of the 1990s, the school suddenly started to enrol fewer children of middle-class parents living and working in the neighbour-hood and more pupils from immigrant families and poor households. To explain this abrupt change, staff members did not only point at demographic changes in the neighbourhood and in Leuven at large, but also at the relocation of a couple of hospitals and other big employers out of the neighbourhood. In the 1990s, the school also initiated special OKAN-classes for foreign language newcomers. In the very beginning, these OKAN-classes mainly attracted children of foreign professors and other highly educated expats, but asylum seekers and refugees gradually became the largest group. As these changes took place, white middle-class parents started to avoid the school. In the terminology of the interviewees, this white flight was followed quickly by a black flight of middle-class pupils with foreign roots (see Peleman, 1998).

> When I look into the numbers, then they have made the change from 35 per cent of target group pupils [eligible for a school grant, not speaking Dutch at home or from a mother without secondary school diploma) to 60 or 70 per cent of target group pupils in two years' time.
>
> (Former school principal)

Multilevel research by Agirdag et al. (2016) demonstrates that multicultural prac-tices are taken up more slowly in Catholic schools such as Mater Dei than in state schools and that ethnic minority teachers report higher levels of multicultural con-tent integration than native white teachers. As such, it is essential to underline that nearly all members of the school staff have Belgo-Belgian roots.

Initially, part of the native, white teaching staff at school struggled to serve the superdiverse group of children and parents. Many school principals and teachers testified that they were anxious about what was happening and admitted that they did not really know how to deal with the changes. In fact, many white professionals acknowledged that they missed the expertise to teach a superdi-verse group of pupils and that they lacked the communication skills to interact with their parents. Because of this, many teachers also opposed the changes that were taking place:

Teachers have opposed the large presence of allochthons for a long time. They really thought that they could not get any further with them.

(Former school principal)

I had the feeling that a gigantic gap had grown between the expectations of the teachers and the reality in which they had to work. (. . .) Some people also admitted that very honestly. 'We do not know any more what to do. We simply do not understand it'.

(Former school principal)

For many years, the staff room at Mater Dei was marked by a conflict between a group of teachers in favour of a radical transformation of the school and another group seeking to preserve the status quo. While the former advocated fundamental changes in the pedagogical project of the school, the communication strategies with parents, the collection of school fees, the role of parents in the school and the cooperation with neighbourhood organizations, the latter argued that it was the responsibility of the school to integrate children of underprivileged and marginalized communities as well as possible in the existing education model. There was a strong tension between both groups:

You have children from double income families who stayed in Leuven after their studies [at university]. You have the original poor groups from Leuven. (. . .) And then you have a continuous inflow of new children. Most of them do not speak Dutch. And that makes such a gigantic mix that the average child does not exist anymore. In a village school, you can allow yourself to work for the average child or parent. And you will reach a very big group. But I understood very quickly that you ignore a lot of people if you do the same at Mater Dei. You have to think and work and communicate in a completely different way.

(Former school principal)

To cling to the past, for one or another reason. In the knowledge that the time for that had gone, but cling to it nonetheless. And the need to . . . The context has changed, the only way to deal with it is to change as well. Those two basic positions were very polarized after three or four years.

(Former school principal)

It took many years before the teaching staff, the school principals and the board of directors had developed a shared vision for the school in a superdiverse context. Eventually, the more conservative forces joined the proponents of more radical transformation. In the resulting transformation process, many issues relating to economic redistribution, cultural recognition and political representation were discussed and resolved. They ranged from the languages being used on the playground and the way birthdays were celebrated in the classrooms to the recuperation of unpaid school fees and the participation of parents in the decision-making process at school. Members of the school staff and the board of directors of Mater Dei invested a lot of their personal time and energy to address these issues. As such, we consider many of the practices of these professionals to be acts of solidarity. Some of these solidarities developed within the classrooms and school walls. They are the focus of the next section. Other forms of solidarity necessitated actions outside the school. They will be discussed in the section thereafter.

Solidarity inside the school

Given the fact that so many children do not have Dutch as their mother tongue, the staff of Mater Dei reconsidered its pedagogical inspirations. Some teachers followed courses on multicultural pedagogy and adapted it to the specific school context. They experimented with methods to use music and other arts as tools for language acquisition. New textbooks to teach Dutch were purchased as well. According to the most recent inspection report of the Flemish government, these strategies were also successful. The inspection report even states that 'the school fell back on a sophisticated language policy to implement present-day language proficiency education in an intensive and exemplary manner' (Flemish Ministry of Education, 2014, p. 4).

Another important change inside the classrooms concerns the development of in-class differentiation. To accommodate multiple learning speeds and preferences, the traditional, classical approaches have been replaced by group work and attention for the needs of each student. This is also materialized in the computer, media and reading corners of every class. While in-class differentiation has become common practice in primary and pre-primary schools all over Flanders (Engels et al., 2014), its early adoption in Mater Dei demonstrates the school's ambition to provide a superdiverse group of pupils with good education:

> Not everyone has the same capacities and chances. We accept those differences and look for the potentialities. It is crucial that all children remain involved and progress at their own speed. To facilitate this, we differentiate our classes and provide individual care whenever needed.
>
> (Mater Dei, 2014)

Other forms of solidarity enacted inside the classrooms act as a testimony of respect for cultural differences. While the mission statement of the school refers unambiguously to the Gospel and underlines that the school observes Christian or Catholic holidays such as Christmas, Easter, Saint Nicolas and Carnival, it also emphasizes that every pupil has the right to respect, recognition and appreciation for his or her being different. This vision is also put into practice. In the classes, teachers leave a lot of room for news from other countries and elements of other cultures. One day, for instance, when we were undertaking participatory observation in one of the pre-primary classes, a girl with Nepalese roots celebrated her birthday. At the start of the day, the teacher called all children to join in a circle.

She asked all of them how they celebrated birthdays at home. Afterwards, she talked about the birthday of the Nepalese girl:

> You're wearing such a nice dress! Is that for your birthday? Is that a dress from Nepal? What kind of clothes do Nepalese people wear? Have you been there often?
>
> (Teacher)

Without doubt, these in-class changes require a lot of effort from the side of the teachers. Teaching for the average student requires much less preparation than catering for the needs of each student individually. For teachers, it is also much more convenient to use the same methods and textbooks year after year, just like it is easier to celebrate all birthdays in the same way. Thinking about the efforts needed to accommodate cultural differences and to cater for different learning speeds and methods, many interviewees argued that teachers received the same salary in all schools, but that working in a school like Mater Dei asked a lot from them. As teachers share so much of their time and energy, we consider their attempts to reform pedagogical strategies, to install in-class differentiation and to accommodate cultural differences to be acts of solidarity.

Acts of solidarity are not only nurtured at the scale of the classroom, but also at the scale of the school. At this scale, practices of solidarity have been set up to deal with issues of cultural recognition, political participation and economic redistribution. In terms of cultural recognition, different school actors have learnt to accommodate the multilingualism of children and parents. The underlying rationale, as expressed by the school principal of the pre-primary school, is that 'you can't deal with all parents in the same way because of background, culture, language . . .'. In official letters, the school uses a simplified Dutch and pictograms. In more informal conversations with parents, the professionals at school adopt different communication strategies, signs, postures and registers of French, English and Dutch.

Initiatives to increase the involvement of parents in school affairs constitute a second trajectory of change within the school. They can be seen as efforts to increase the political participation of parents without a white middle-class background. Once the school population had changed significantly, different actors at school lamented the fact that the traditional parents' council was mostly made up of white middle-class parents. First, attempts to create a cultural mix within the parents' council failed because newcomers struggled to find their place within the already established norms, values and power relations. Hence, the school started to experiment with alternative forms of parents' participation. On the one hand, this is visible in an emphasis on presence and proximity. The drop-off and pick-up procedures have been re-organized in such a way that they allow for informal chats with the teaching staff. Sometimes, the school principals also greet the children and the parents at the school gates. By being accessible, teachers and school principals allow children and parents to approach them with small and big problems. In addition, an employee of the local community organization has also been invited to serve as a representative of the local community in the school council.

As a spokesperson of the disadvantaged parents she meets on a daily basis, she has raised a number of important issues already:

> I am a member of the school council. I represent the people of our community organization. And if we discuss issues such as the cost of school trips, I have a say. Often, the parents have to pay for the bus. Then we ask to take the character of our neighbourhood into account.
>
> (Employee of the local community
> organization involved in the school)

Rather than ethnic, cultural and linguistic diversity, many interviewees consider poverty to be the main challenge at school. Often, teachers are confronted with empty lunch boxes. The school also struggles with a huge backlog of unpaid school bills. Obviously, the school staff are aware of the financial limitations of many parents. In Flanders, the law stipulates that schools cannot ask more than 45 or 85 euro a year for activities and materials in the pre-primary and primary school, respectively. Over the six years at primary school, an additional 410 euros can also be demanded for multi-day trips. Even before this law came into place, the school made great effort to limit the costs of trips and activities. The head of the primary school told us, for instance, that 'teachers look everywhere for excursions which are free or very cheap'. Nonetheless, many respondents also felt that the scale of the school was not the right scale to address issues of economic redistribution. As such, we turn to the scales of the neighbourhood, the city and the Flemish region in the next section.

Striving for solidarity beyond the school

For many interviewees, it is clear that problems and challenges in the here and now of the daily functioning of the school are strongly interwoven with long-term structural processes related to poverty and the provision of social benefits and services. As such, some stakeholders at Mater Dei underlined that place-bound mechanisms of solidarity within the school walls are not sufficient to realize equal opportunities for all pupils. Taking into account that structural mechanisms of exclusion operate beyond the territory of the school, they argue that their efforts to enact solidarity at school need to be sustained by practices of solidarity outside the school:

> That small school is only such a small piece of the problem (. . .) It is impossible for me to make a difference in all these domains. And that is a frustration. I mean, that is a challenge.
>
> (Former school principal)

> You can play with the kids and provide democratic meals, but something needs to change structurally as well and that is only possible through policy interventions.
>
> (Coordinator of the local community
> organization involved in the school)

The neighbourhood is a first scale at which solidarity beyond the school is called for. Over the years, several members of the Mater Dei staff made a lot of efforts to develop a network of neighbourhood organizations in which the school was the central node. In the words of one of the school principals, the cooperation with community organizations, community centres and a local health centre makes it 'difficult to separate inside and outside at school'. This inside-outside blurring works in two ways. On the one hand, the school refers parents with specific questions and problems to these organizations. At the same time, they also invite them to take up a role in the educational project of the school. By way of example, one of the local community organizations sets up homework classes and lunch activities for the pupils of Mater Dei.

> In the struggle against the unequal opportunities experienced by our children, the school and the staff do not stand alone. With organizations from the immediate neighbourhood, (. . .) excellent contacts are being maintained.
>
> (Mater Dei, 2011)

While many employees of the school had the feeling that issues of cultural recognition and political representation could be discussed and addressed inside the school, many felt that issues of economic redistribution required solutions at larger spatial scales. In the past, when many children from dual-earning families were enrolled at Mater Dei, the school earnt money by providing warm lunches and before and after school care. With this money, the school could turn a blind eye to parents who struggled to pay the lunches, the school bills or the school trips of their children. Yet, once the socio-economic background of the school population had changed radically, and once the lunch and care services had been outsourced to private agencies, such informal practices of redistribution had become much more difficult to set up. In fact, one of the former school principals explained that the school had thousands of euros in arrears at a particular moment. Once children from disadvantaged parents made up the majority of the school population, many stakeholders claimed, it was hard to organize economic redistribution within the school walls:

> There is an increase of unpaid bills. The financial difficulties. . . The financial means generated by the school itself. . . There is not enough money. You're paying the price of the poverty as well.
>
> (Former school principal)

> In the past, when there was before and after school care, we could organize more solidarity.
>
> (Former school principal)

Staff members of Mater Dei made great efforts to minimize the financial burden on the parents. The board of directors of the school instigated the city council to allow schools to visit city museums at no cost. The chairperson of the board also became a strong advocate of social mix to be organized at the urban scale. In meetings with other chairpersons of the board of directors of other Catholic schools, he argued that school segregation had become a problem in Leuven. In his eyes, the concentration of pupils with an immigrant background and/or a lower socio-economic status in 'concentration schools' such as Mater Dei was not only problematic because of its impact upon the financial means of the school, but also because it limited the chances that children with different ethnic and cultural backgrounds and socio-economic profiles would encounter each other, both inside and outside his own school.

Despite efforts by the city council to attract children from different socio-economic backgrounds to each school, Leuven is still marked by high levels of school segregation, however. In the absence of a school catchment system, many middle-class parents choose to enrol their children in schools with a low proportion of pupils from poor and immigrant families. Some interviewees even indicated that a couple of schools were eager to refer immigrant children to Mater Dei in the past. They also doubted whether all schools were willing to go through the kind of learning trajectory Mater Dei had gone through and to take the kind of measures that Mater Dei had taken to deal with poverty and to accommodate superdiversity. In the opinion of many stakeholders, the effects of solidarity with the Mater Dei pupils would be limited as long as other schools did not develop the same kinds of practices of solidarity with their children. As such, they became fierce promotors of solidarity between schools in Leuven.

A last scale to be considered is the scale of the region. In subsequent rounds of Belgian state restructuring, education has become a responsibility of the Flemish government. In Flanders, schools like Mater Dei are able to provide good quality education because they receive more government funding per student than elite schools do, both for materials and teaching staff. Recently, the minister of education has started discussions, however, to reconsider this redistribution of resources between schools. Since the school population – and therefore also the government funding – of Mater Dei has been decreasing for a number of consecutive years, there are serious concerns that the school will not be able to continue doing what they are currently doing because of a lack of financial resources. Basically, actors at Mater Dei assert that they can only be in solidarity with the children and parents at school because of the structural forms of solidarity the Flemish government has set up between schools (see Schuermans et al., 2014).

Sources of solidarity

For many people at Mater Dei, practices of solidarity are part of a struggle for the right to good education. Different actors join this struggle for different reasons. Confronted with the structural disadvantages of the children, some find inspiration in their own youth spent in poverty. Others refer to their socialist upbringing or their Christian faith to explain their commitment. While the representative of the pupil guidance centre clarified that his efforts are inspired by his social orientation, an old nun who used to be involved in the school explained that she sees her work with the underprivileged children as an apostolic mission. In spite of these very different motives, and in spite of the disagreements between different partners, the joint struggle against educational injustices is marked by a strong feeling that 'we are rolling it together', as one of the interviewees put it:

> [The chairperson of the board of directors] joined Mater Dei because he was so shocked by that injustice. (. . .) He knew about my struggle and he joined it almost unconditionally.
>
> (Former school principal)

> It is more than about being shocked by the injustice. It is about a right for everyone. Everyone has a right to good education.
>
> (Current school principal)

Both for people who self-identify as Christians and for those who do not, the Christian roots of the school support their solidarities. In interviews, it was often mentioned that the school had been established for orphans from its very outset. The homepage on the school website also cites the Book of Matthew to explain its 'commitment to the weakest, the most vulnerable, the poorest in society'. While such sentences remain hollow phrases in many other Catholic schools opting for a more elitist trajectory, Christian norms and values provide an institutional embedding of solidarity at Mater Dei, even if they are not shared by many non-believers:

> As a Catholic school, we want to live the central values of love and care and transmit them to our children through our daily efforts and our concrete acts. Our Christianity remains a lively source of inspiration for our commitment to the weakest, the most vulnerable, the poorest in society. Indeed: 'Whatever you did for one of the least of these brothers and sisters of mine, you did for me' (Mat, 25: 40).
>
> (Mater Dei, 2014)

Naturally, it may not be forgotten that many stakeholders are much less progressive than the ones cited above. According to some interviewees, many teachers were conservative, nostalgic or sometimes even racist. It cost a lot of effort to convince them to adopt the new vision and the underlying pedagogic principles. In this process, encounter was considered to be a crucial source of solidarity. To get more familiar with the daily life of the children and their parents, the school principals organized workshops to get to know the neighbourhood. They also arranged staff meetings in the buildings of a local community organization. The underlying rationale was that staff members would start to show more understanding for the situation of the pupils if they would be more aware of their living circumstances. Through an immersion in the lifeworld of the pupils and their parents, the school principals and the board of directors hoped for more insight, more respect and more solidarity. While this did not always work out, it was interesting to see that encounter was considered to be a source of solidarity in the absence of a collective commitment to a common struggle or shared norms and values:

I needed to invest very much in teachers with regards to that change. I had to fight their resistance. That has cost a lot of energy.

(Former school principal)

We did several activities to familiarize ourselves with the neighbourhood. On Friday evenings, we tried to go to the pub together. If we went for a beer, it was in the neighbourhood. We also organized a workshop where we sent teachers to all kinds of neighbourhood organizations.

(Former school principal)

By analysing the role of Catholic norms and values and the struggle for equal educational opportunities for all, one comes to understand both the potential and the limits of interpersonal encounters in the development of solidarity in superdiversity. On the one hand, the Mater Dei case study shows that many encounters across difference generate defensive attitudes, reactionary discourses and practices of 'desolidarization'. On the other hand, the case also demonstrates that encounters across difference are strongly mediated by people's biographies (e.g. growing up in poverty, a socialist upbringing, etc.) and beliefs (e.g. Christian values, socialist orientations, etc.). While much of the geographical and socio-logical literature on encounter almost seems to suggest that it is sufficient to gather people of different backgrounds in one place to achieve incremental change in values, discourses and/or practices, our analysis of different sources of solidarity, therefore, demonstrates that there is certainly no direct link between the here-and-now of the encounter and the there-and-then effect under the form of solidarity.

Conclusion

This chapter has aimed to disentangle the development of place-based solidari-ties in a formal schooling environment that has rapidly become superdiverse. In contrast to most of the sociological and geographical literature on superdiverse schools, we did not focus on the meaningful interactions between pupils and par-ents of different socio-economic, ethnic and cultural backgrounds, but on the way in which teachers, principals and other school personnel encounter students and parents who are not like them. Drawing on participatory observations, document analysis and in-depth interviews with personnel of the school, we are able to come back to the three points made in the introduction to this chapter. As such, we draw conclusions about 1) the effects of encounter; 2) the place-boundedness of these effects; and 3) the relationship between the encounter, on the one hand, and the effects, on the other.

First, our analysis clarifies that encounters with a superdiverse group of stu-dents do not only impact upon the way teachers and other school personnel perceive the world and talk about it, but also upon the way they act upon it. Many interviewees indicated, indeed, that they had not only reconsidered their values and discourses about children from different backgrounds, but that they had also changed their do's and don'ts in the school over time. Inspired by the conceptual framework outlined in the introduction to this volume, we consider the time and the energy members of the school staff share with a group of children and parents of various socio-cultural and socio-economic backgrounds to resolve issues of cultural recognition, political participation and economic redistribution as practices of solidarity (see Oosterlynck et al., 2016). Examples of cultural recognition range from the way in which birthdays are being celebrated in the classroom to the use of pictograms in written communication with parents. Political participation at the school level is being transformed by the involvement of a community organization in the school council and the increased emphasis on professional presence and proximity. Economic redistribution is being pursued through claims to make the city museums free of charge for school groups and by turning a blind eye to unpaid school bills as long as that was possible.

In relation to the second point, it is clear that many of the solidarities discussed in this chapter are place-based, but not place-bound. While many teachers, principals and other professionals working at the school indicate that many issues of political participation and cultural recognition can be resolved within the territory of the school, they also emphasize that the financial limits of the school are quickly reached when they try to address issues of economic redistribution. Hence, different stakeholders at Mater Dei have turned the school into a node in a network of neighbourhood organizations and have called for mechanisms of redistribution between schools, both inside and outside Leuven. This shows that proximity based solidarities nurtured by encounters in the here and now can transcend distances through professional interventions.

As a nuance to much of the sociological and geographical literature on contact and encounter, and in line with the third point made in the introduction, it is clear, however, that there is no direct, causal relationship between the interpersonal interactions at school and the place-based practices of solidarity discussed above. While some professionals consider their daily encounters across lines of class, ethnicity and culture to be calls to be acted upon, other colleagues do not. Even though progressive forces within the school count on intercultural encounters outside the school territory as a means to nurture solidarity among more conservative colleagues, such encounters alone are not sufficient to reach these positive effects. Instead, it needs to be acknowledged that many of the encounters and solidarities at Mater Dei are strongly mediated by personal histories, pre-existing norms and values and struggles for social justice.

Acknowledgements

This chapter presents the results of a case study conducted within the interdisciplinary DieGem project on solidarity in diversity in Flanders. We would like to thank the Institute for the Promotion of Innovation through Science and Technology in Flanders (IWT-Vlaanderen) for their financial support and Stijn Oosterlynck and Maarten Loopmans for their valuable comments on earlier versions of this chapter. Obviously, the usual disclaimer applies.

Bibliography

Agentschap voor Onderwijsdiensten (2015). *Statistieken Leerlingenkenmerken [Statistics pupil characteristics]*. Available online at www.ond.vlaanderen.be/wegwijs/agodi/cijfermateriaal/ leerlingenkenmerken/, last consulted 7 January 2016.

Agirdag, O., Merry, M. S., & Van Houtte, M. (2016). Teachers' understanding of multicultural education and the correlates of multicultural content integration in Flanders. *Education and Urban Society, 48*(6), 556–582.

Agirdag, O., Van Houtte, M., & Van Avermaet, P. (2013). School segregation and self-fulfilling prophecies as determinants of academic achievement in Flanders. In S. De Groof & M. Elchardus (Eds.), *Early school leaving and youth unemployment* (pp. 46–74). Amsterdam Lannoo: Amsterdam University Press.

Allport, G. (1954). *The nature of prejudice*. Cambridge, MA: Addison-Wesley.

Amin, A. (2002). Ethnicity and the multicultural city: Living with diversity. *Environment and Planning A, 34*(6), 959–980.

Amin, A. (2012). *Land of strangers*. Cambridge: Polity Press.

Araújo, M. (2016). A very 'prudent integration': White flight, school segregation and the depoliticization of (anti-) racism. *Race, Ethnicity and Education, 19*(2), 300–323.

Arnaut, K. (2013). Super-diversity: Elements of an emerging perspective. *Diversities, 14*(2), 1–15.

Binnie, J., Holloway, J., Millington, S., & Young, C. (2006). *Cosmopolitan urbanism*. New York: Routledge.

Blau, P. M. (1977). *Inequality and heterogeneity: A primitive theory of social structure*. New York: Free Press.

Brooks, J. S. (2012). *Black school, White school: Racism and educational (mis)leadership*. New York: Teachers College Press.

Collins, D., & Coleman, T. (2008). Social geographies of education: Looking within, and beyond, school boundaries. *Geography Compass, 2*(1), 281–299.

Cook, V. A., & Hemming, P. J. (2011). Education spaces: Embodied dimensions and dynamics. *Social & Cultural Geography, 12*(1), 1–8.

Danhier, J., Jacobs, D., Devleeshouwer, P., Martin, E., & Alarcon-Henriguez, A. (2014). *Naar kwaliteitsscholen voor iedereen?: Analyse van de resultaten van het PISA 2012 onderzoek in Vlaanderen en in de Federatie Wallonië-Brussel [Quality schools for everyone? Analysis of the 2012 PISA research results in Flanders and the Wallonia-Brussels Federation]*. Brussels: Koning Boudewijnstichting.

Dejaeghere, Y., Hooghe, M., & Claes, E. (2012). Do ethnically diverse schools reduce ethnocentrism? A two-year panel study among majority group late adolescents in Belgian schools. *International Journal of Intercultural Interrelations, 36*(1), 108–117.

D'hondt, F., Van Praag, L., Stevens, P. A., & Van Houtte, M. (2015). Do attitudes toward school influence the underachievement of Turkish and Moroccan minority students in Flanders? The attitude-achievement paradox revisited. *Comparative Education Review, 59*(2), 332–354.

Engels, N., Struyven, K., & Coubergs, C. (2014). Leerkrachten uitgedaagd: omgaan met diversiteit [Teachers challenged: Dealing with diversity]. In I. Nicaise, D. Kavadias, B. Spruyt, & M. Van Houtte, M. (Eds.), *Het onderwijsdebat. Waarom de hervorming van het secundair broodnodig is [The educational debate. Why the reform of secondary education is highly necessary]* (pp. 211–240]. Berchem: EPO.

Flemish Ministry of Education. (2014). *Verslag over de opvolgingsdoorlichting van Vrije Lagere School – Mater Dei te Leuven [Report about the successive inspection of the Mater Dei primary school in Leuven]*. Brussels: onderwijsinspectie.

Flemish Ministry of Education. (2015). *Voorpublicatie Statistisch Jaarboek 2014–15.* www.ond.vlaanderen.be/onderwijsstatistieken/2014-2015/statistischjaarboek2014-2015/ publicatiestatistischjaarboek2014-2015.htm, last consulted 7 January 2015.

Geldof, D. (2013). *Superdiversiteit. Hoe migratie onze samenleving verandert [Superdiversity. How migration changes our society].* Leuven: Acco.

Hallinan, M., & Williams, R. A. (1989). Interracial friendship choices in secondary schools. *American Sociological Review, 54*(1), 67–78.

Havermans, N., & Hooghe, M. (2011). *Kerkpraktijk in België: Resultaten van de zondags- telling in oktober 2009.* Leuven: Centrum voor Politicologie.

Hemming, P. J. (2011). Meaningful encounters? Religion and social cohesion in the English primary school. *Social & Cultural Geography, 12*(1), 63–81.

Holloway, S. L., Hubbard, P., Jöns, H., & Pimlott-Wilson, H. (2010). Geographies of education and the significance of children, youth and families. *Progress in Human Geography, 34*(5), 583–600.

Jacobs, D., Rea, A., Teney, C., Callier, L., & Lothaire, S. (2009). *De sociale lift blijft steken. De prestaties van allochtone leerlingen in de Vlaamse Gemeenschap en de Franse Gemeenschap [The social lift comes to a halt. The performance of immigrant pupils in the Flemish community and the French community].* Brussels: Koning Boudewijnstichting.

Kong, L. (2013). Balancing spirituality and secularism, globalism and nationalism: The geographies of identity, integration and citizenship in schools. *Journal of Cultural Geography, 30*(3), 273–307.

Maly, I. E. L., Blommaert, J. M. E., & Yakoub, J. B. (2014). *Superdiversiteit en democratie [Superdiversity and democracy].* Berchem: EPO.

Massey, D. (2004). Geographies of responsibility. *Geografiska Annaler B, 86*(1), 5–18.

Massey, D. (2005). *For space.* London: SAGE.

Mater Dei. (2011). *Aangepast Gelijke Onderwijs Kansen (GOK)-plan [Revised plan equal chances in education].* Leuven: Mater Dei.

Mater Dei. (2014). Website Mater Dei. www.materdei-leuven.be/, last consulted 2 July 2014.

Moody, J. (2001). Race, school integration, and friendship segregation in America. *American Journal of Sociology, 107*(3), 679–716.

Mouw, T., & Entwisle, B. (2006). Residential segregation and interracial friendship in schools. *American Journal of Sociology, 112*(2), 394–441.

Neal, S., & Vincent, C. (2013). Multiculture, middle class competencies and friendship practices in super-diverse geographies. *Social & Cultural Geography, 14*(8), 909–929.

Noble, G. (2013). Cosmopolitan habits: The capacities and habitats of intercultural con- viviality. *Body & Society, 19*(2&3), 162–185.

Nowicka, M., & Vertovec, S. (2014). Comparing convivialities: Dreams and realities of living-with-difference. *European Journal of Cultural Studies, 17*(4), 341–356.

Oosterlynck, S., Loopmans, M., Schuermans, N., Vandenabeele, J., & Zemni, S. (2016). Putting flesh to the bone: Looking for solidarity in diversity, here and now. *Ethnic and Racial Studies, 39*(5), 764–782.

Peleman, K. (1998). *Concentratiescholen of buurtscholen? [Concentration schools or neighbourhood schools?].* Planologisch Nieuws, *17*(4), 307–316.

Pulinx, R., Van Avermaet, P., & Agirdag, O. (2015). Silencing linguistic diversity: The extent, the determinants and consequences of the monolingual beliefs of Flemish teachers. *International Journal of Bilingual Education and Bilingualism,* doi: 10.1080/13670050.2015.1102860.

Quillian, L., & Campbell, M. E. (2003). Beyond black and white: The present and future of multiracial friendship segregation. *American Sociological Review, 68*(4), 540–566.

Santagati, M. (2015). Researching integration in multiethnic Italian schools. A sociological review on educational inequalities. *Italian Journal of Sociology of Education, 7*(3), 294–334.

Saunders, D. (2010). *Arrival city: How the largest migration in history reshaped our world*. New York: Pantheon Books.

Schuermans, N. (2014). Geography textbooks and the reproduction of a racist and ethnocentric world view among young people in Flanders. *Belgeo*, https://belgeo.revues.org/11594?lang=fr.

Schuermans, N. (2016). Enclave urbanism as telescopic urbanism? Encounters of middle class whites in Cape Town. *Cities*, 59, 183–192.

Schuermans, N., Debruyne, P., & Jans, M. (2014). Gelijke onderwijskansen kosten bloed, zweet en vooral veel geld [Equal opportunities in education require, blood, sweat and a lot of money]. *De Morgen*, retrieved from www.demorgen.be/binnenland/gelijke-onderwijskansen-kosten-bloed-zweet-en-vooral-veel-geld-b80165c0/.

Spruyt, B. (2008). Ongelijkheid en segregatie in het onderwijslandschap: effecten op etnocentrisme [Inequality and segregation in the educational landscape: Effects on ethnocentrism]. *Tijdschrift voor Sociologie, 29*(1), 60–89.

Stad Leuven (2015). *Leuven in cijfers [Leuven in numbers]*. Retrieved from www.leuven.be/bestuur/leuven-in-cijfers/bevolking/nationaliteit/.

Stjernø, S. (2004). *Solidarity in Europe: The history of an idea*. Cambridge: Cambridge University Press

Thijs, J., & Verkuyten, M. (2014). School ethnic diversity and students' interethnic relations. *British Journal of Educational Psychology, 84*(1), 1–21.

Uytterhoeven, R., & Van den Heuvel, H. (1994). *Twee eeuwen Zusters van Liefde [Two centuries Sisters of Love]*. Leuven: Zusters van Liefde Leuven.

Valentine, G. (2008). Living with difference: Reflections on geographies of encounter. *Progress in Human Geography, 32*(3), 323–337.

Van Avermaet, P., & Sierens, S. (2010). *Diversiteit is de norm. Er mee leren omgaan de uitdaging. Een referentiekader voor omgaan met diversiteit in onderwijs [Diversity is the norm. Learning to deal with it the challenge. A framework to deal with diversity in education]*. In D. De Coen et al. (Eds.), *Handboek beleidsvoerend vermogen* (pp. 1–48). Brussel: Politeia.

Van den Broucke, S., Noppe, J., Stuyck, K., Buysschaert, P., Doyen, G., & Wets, J. (2015). *Vlaamse Migratie- en Integratiemonitor 2015 [Flemish Migration and Integration Monitor 2015]*. Antwerpen/Brussel: Steunpunt Inburgering en Integratie/Agentschap Binnenlands Bestuur.

Van Houtte, M., & Stevens, P. A. J. (2009). School ethnic composition and students' integration outside and inside schools in Belgium. *Sociology of Education, 82*(2), 217–239.

Vertovec, S. (2007). Super-diversity and its implications. *Ethnic and Racial Studies, 30*(6), 1024–1054.

Wessendorf, S. (2013). Commonplace diversity and the 'ethos of mixing': Perceptions of difference in a London neighbourhood. *Identities, 20*(4), 407–422.

Wilson, H. F. (2014). Multicultural learning: Parent encounters with difference in a Birmingham primary school. *Transactions of the Institute of British Geographers, 39*(1), 102–114.

Wilson, H. F. (2016). On geography and encounter: Bodies, borders, and difference. *Progress in Human Geography*, doi: 10.1177/0309132516645958.

Wise, A., & Velayutham, S. (2009). *Everyday multiculturalism*. Basingstoke: Palgrave Macmillan.

4 Building coalitions

Solidarities, friendships and tackling inequality

Helen F. Wilson

1. Introduction

> A sheet of A2 flip-chart paper is pulled off to reveal a T diagram with the words 'advocacy' and 'coalition' scrawled across the top in green pen. 'Now, it's always good to reflect on the difference between the two. . .' Nina, the chair of the meeting, raises her pen to the paper and makes a spot under advocacy. She pauses and looks over her shoulder at the group of people gathered around the room. 'Aiding a particular cause?' somebody offers. 'Arguing for something or presenting a specific case?' These go up on the board. The pen moves across to coalition. There is another pause. A youth worker, who is sat on the floor, offers a definition: 'an alliance across different groups and causes'. 'Excellent!,' Nina punches the air, 'and that's what we're focused on – the development of coalitions across different groups and causes to create more just communities!'
>
> (Observation, USA, 2015)

This exchange marked the end of a monthly chapter meeting for The Community Leadership Network (CLN).[1] Presented as an opportunity for staff and volunteers to reconnect with each other, refresh training, reflect on goals and share information about ongoing projects, the session was also a sociable event. Food was shared, family and friends were brought along, and personal news was exchanged. CLN is a non-profit, international training organization with a broad remit for work on inequality. Founded in the US 30 years ago to tackle inter-ethnic conflict on college campuses, it has since developed specialties in conflict resolution and mediation, dialogue programmes, anti-violence work, and workshops focused on prejudice and discrimination. It has a network of more than 50 branches across North America and Europe.

Using CLN as a case study, this chapter concerns the labour, resources and connections that go into building solidarities and takes CLN's emphasis on the development of 'coalitions' as its starting point. Rather than focusing on the outcomes, the chapter prioritizes the efforts that go into achieving social change (see for example Alderman et al., 2013) and focuses specifically on the interpersonal and affective relationships that lie at its heart. It does two things.

First, it connects with recent work on solidarity that has concerned the emotional labour, organizational capacities, knowledge, personal commitments and sacrifice that go into building transformative relations of various kinds (Fariás, 2016; Featherstone, 2010; Oosterlynck et al., 2016; Vasudevan, 2015). Second, it responds to calls for a rethinking of how we understand social activism more broadly, to better prioritize activisms that are 'small-scale, personal and quotidian' (Horton & Kraftl, 2009, p. 14).

The chapter adds to the work outlined above through an examination of the links between solidarity and friendship and makes three contributions. First, in noting how the *geographies* of friendship still remain relatively under-theorized (Bowlby, 2011; Bunnell et al., 2012; Conradson & Latham, 2005; Hall & Jayne, 2016; Wilkinson, 2014), it demonstrates how a focus on friendship can offer novel insights into the assembling of solidarities that cut across social differences, interests and places, while also drawing attention to what else is produced in the building of coalitions (Askins, 2015). Second, by attending to the spaces in which friendships are built, the chapter builds on debates surrounding the geographies of encounter (Darling & Wilson, 2016; Schuermans, 2013; Valentine, 2008). These debates have not only considered how people negotiate difference in spaces of interaction, but have asked what potential such spaces hold for catalysing change, both pedagogically and politically (Askins & Pain, 2011; Halvorsen, 2015; Ince, 2015; Wilson, 2016). Finally, by focusing on solidarities and friendships in the context of a largely voluntary organization, the chapter adds to work on the 'enlivened geographies' of volunteering (Smith et al., 2010; see also Conradson, 2003; Fyfe & Milligan, 2003). This is important given that so much of the work on solidarity has focused on movements, organizations and individual actions that are largely sustained through voluntary practices. Recent work on volunteering has underlined the need to understand the ordinary processes, emotions and situated relationships that are often overlooked in social policy so as to make important links between practice agendas and what happens in the taking-place of voluntary work (Darling, 2011). As I argue, understanding the motivations for volunteering is crucial, whether place-based, biographical or a sense of efficacy, but so too is an understanding of what *sustains* voluntary practice, and this forms a key part of the chapter's focus (Mills, 2013; Smith et al., 2010; Warren, 2014).

The chapter draws on multi-sited fieldwork, which involved in-depth interviews with members of CLN, along with participants and partner organizations who have worked or trained with the charity in both Europe and the US. Interviews focused on the mobility of programmes, knowledge and resources; the taking-place of anti-violence work, dialogue and training; and the wider impacts and experiences of CLN's various projects and programmes. Participant diaries were also completed, allowing participants to record and reflect on their experiences in the days and weeks after CLN workshops to provide further material for discussion during interviews. Observant participation of meetings, training and programme events were combined with analysis of training material, as well as local and national policy documents so as to provide context for current priorities in matters of discrimination and social conflict. This research thus combines an interest in the mobility of ideas, people and resources, with a focus on the spaces in which mediation and anti-violence work takes place (Wilson, 2012, 2013a, 2014).

Following an overview of recent but separate debates on friendship and solidarity, the third section begins with a space of encounter – a CLN annual general meeting (AGM) – as an introduction to the relationships at the heart of the organization. Sections 4 and 5 then consider how emotional connections are built through engaging in struggles against inequality. In placing emphasis on what emotions *do* (Ahmed, 2004), the chapter notes how personal relationships become instrumental to – and inseparable from – the mobilization of action. As I argue, while these narratives humanize and complicate accounts of social activism, they also thoroughly underline the entangled nature of solidarity and friendship. By building on these complicated narratives and reflecting on how personal connections are maintained, Section 6 highlights how coalitions are built across different causes and place-based struggles – whether anti-racist agendas, LGBTQ issues, indigenous rights or any number of others. Finally, I finish with some reflections on what can be gained from paying attention to friendship in matters of solidarity, emphasizing non-linear accounts of activism and the less spectacular ways in which activism gets done.

2. Solidarities, friendships and affective attachments

As geographers have recently highlighted, friendship, and its embodied, affective and emotive dimensions of support, can play a significant role in social transformation and offer novel readings of place and co-existence (Bunnell et al., 2012; Cronin, 2015b). Such assertions feed into a wider body of scholarship that has demonstrated how all forms of 'close affective encounter' and intimacy (Oswin & Olund, 2010, p. 62) are constitutive of a variety of economic, political and geopolitical practices (Pain & Staeheli, 2014; Povinelli, 2006; Smith, 2016; Vasudevan, 2015; Wilson, 2016). For example, attention to friendship in the context of diversity can shape how we understand alternative narratives of community and belonging, and can rupture and undermine negative discourses concerning the future of plural cities (Bunnell et al., 2012; Kathiravelu, 2013; Neal & Vincent, 2013; Vertovec, 2015; Wilson, 2013b). Rather than reducing friendship to a set of private, interpersonal relationships, friendship thus becomes an important lens through which to understand the wider 'outcomes of metropolitan life' and collective struggle (Kathiravelu, 2013, p. 7; Gandhi, 2005).

Specifically, by concerning affective attachments that lie outside of familial relations and sexual coupledom (Cronin, 2015a; Povinelli, 2006), a focus on friendship necessarily extends debates on intimate life (Berlant, 2000; Hall & Jayne, 2016; Wilkinson, 2014). For instance, Wilkinson's focus on intimacy beyond 'the totalising logic of the couple' (2014, p. 2460) underlines how friendship becomes significant to the construction of home spaces for single people. In so doing, Wilkinson challenges heteronormative temporalities, and highlights forms of care that are regularly overlooked in favour of nuclear family attachments (Ahmed, 2010; Halberstam, 2005). It is not surprising then, that friendship has been widely considered for its radical and counter-normative possibilities (Derrida, 2005; Foucault, 1997; Gandhi, 2005), and that these considerations have challenged the tendency to limit friendship to questions of homophily (see Cronin, 2014 for a discussion). Whether it arises from situations of 'throwntogetherness' (Massey, 2005; Neal & Vincent, 2013; Wilson, 2013a) or whether it is purposefully sought out, friendship is considered for its potential to refuse exclusionary communities of belonging and to disrupt alignments along axes of filiation (Gandhi, 2005, p. 10; Askins, 2016; Derrida, 2005). It can thus rupture social orders, encourage one to live otherwise and bridge differences of class, sexuality, religion, race and intellect.

It is at the intersection between debates on intimacy, friendship and living with difference that this chapter is positioned. In focusing on the interconnections between friendship and solidarity, I also acknowledge what makes them distinct, to note that while friendships are clearly transformative and have the potential to challenge inequalities, they are not defined by these goals (Desai & Killick, 2010). By contrast, solidarities are often forged through struggle and are 'inventive' in that they seek out new ways of 'configuring the political' (Featherstone, 2012). In attending to friendship in this context, I emphasize the often-overlooked relationships that become central to – and emerge out of – transnational collaboration, political struggle and voluntary practices. In this vein, I join scholars that have sought to rethink how we address the affective attachments that shape, and are shaped by, activist spaces (Brown & Pickerill, 2009b).

My account takes it leave from Gandhi's (2005) study of the role of 'unlikely friendships' in late Victorian radicalism. Attending to the minor forms and narratives of anti-imperialism that emerged in Britain during the end of the nineteenth century, Gandhi's account of affective community focuses on the relationships that developed between different, often marginalized groups (such as spiritualists, homosexuals and animal rights activists) and colonized subjects and cultures. Describing them as minor forms of 'crosscultural collaboration between oppressors and oppressed' (ibid., p. 6), her work has questioned what encouraged some individuals within the British Empire to 'betray the claims of possessive nationalism in favour of, solidarity with foreigners, outsiders, alleged inferiors' (ibid., p. 2).

In focusing on anti-colonial actors that reject the binary logics of the imperial culture in which they were situated, Gandhi (2005) highlights how friendship acts as a critical resource for enacting political change and struggle and thus provides an example that is key to thinking about the value of friendship to forms of solidarity. She outlines three starting points for thinking through the *politics* of friendship that are particularly instructive. The first is the radical potential identified by Derrida. As she argues, friendship is understood to be 'the most comprehensive philosophical signifier for all those invisible affective gestures that seek expression outside, if not against, possessive communities of belonging' (ibid., p. 10). Second, to demonstrate the strength of the relationships that she identifies, Gandhi (2005, p. 10) references Forster, who stated that 'If [he] had to choose between betraying [his] country and betraying [his] friend, [he] hoped [he] should have the guts to betray [his] country.' In so doing, she highlights a set of loyalties that make for stronger attachments than other forms of belonging or imagined community. Finally, Gandhi connects these reflections on friendship to Hardt and Negri's (2001) call for new forms of solidarity, when, at the turn of the century, they claimed that 'the time [was] ripe to refuse the ambivalent mantle of citizenship in order to foment a new politics of anti-imperialism, closely attentive to forms of transnational or affiliative solidarity between diffuse groups and individuals' (cited in Gandhi, 2005, p. 10). For Gandhi (2005, p. 26), if friendship is an improvisational politics of co-belonging, which is distinct from the established affective formations of family, fraternity, genealogy and so on, then it can play a significant part in shaping the forms of transnational politics that Hardt and Negri (2001) called for.

It is important to note that not all forms of friendship have such radical potential and that friendships can also have destructive capacities (Smart et al., 2012). Nevertheless, accounts like Gandhi's can make clear contributions to recent scholarship on solidarity, which has emphasized the significance of interpersonal relations, affective attachments and emotional ties, not only to the building of solidarities (Featherstone, 2012; Routledge et al., 2007; Routledge, 2012), but to connecting places and bridging international borders (McFarlane, 2006; Pratt, 2008). This is coupled with an insistence on centring 'subaltern agency' in accounts of political struggle (Featherstone, 2012, p. 247), to acknowledge those individuals and marginalized groups that are rarely accounted for in narratives that often favour key actors, and polished, sanitized stories of social change (Alderman et al., 2013; Horton & Kraftl, 2009). In wanting to hold on to the messiness of political action, it is also necessary to question how forms of solidarity both give rise to, and further sustain friendships, to offer readings that reflect the entangled and mutually constitutive nature of these relations.

Through attending to the personal sacrifices, material resources, aspirations, organizational labour and temporalities that make solidarities possible, scholars such as Featherstone (2012, p. 5) have prioritized the politicized and open-ended construction of solidarities to demonstrate how 'new ways of relating between diverse people and places are formed' *in practice*. Indeed, in line with Featherstone (2012, p. 6), this chapter argues that better attention to the conditions under which solidarities are developed reveals how they can emerge from uneven power relations that cut across social distinctions. In so doing, the chapter necessarily focuses on spaces of encounter as sites of pedagogic and political potential (Wilson, 2016), where trust, mutuality and durable interpersonal relations can develop as central to, and inseparable from, the forms of solidarity and struggle described. As the chapter will argue, these spaces present the opportunity for face-to-face communication, but they also offer a chance to negotiate boundaries, learn about new struggles and enhance motivation, while blurring the boundaries between public and private (Chatterton et al., 2013; Halvorsen, 2015). With this in mind, it is to a space of encounter that I now turn to reflect on the intimacies and affective relations that form a core component of CLN's work.

3. Meetings, support and standing together

> This feels so much like a family reunion. . . and this sense of a family reun-
> ion, is really a manifestation of our core value, which is relationships. . .

This was the Chief Executive Officer's opening line at a CLN AGM in 2014.
The audience of around 50, gathered before her, had travelled from across North
America and Europe to meet, celebrate, exchange information, catch up, reconnect and undertake training. They represented just some of CLN's 50 branches,
and included directors, representatives from partner organizations, workshop
facilitators, and volunteers, who were all involved in the charity's social activism
in one way or another. Some members of the audience had been working with
CLN for more than 28 years, while others were in their first 12 months. For most,
this was the start of their annual reunion and one that was live-streamed for those
of the 'family' who were unable to travel.

The reference to family in this context, hints at the strength of the bonds
between the people in the room, offering an extended sense of intimate life and
supporting academic work that has sought to critique narrow definitions of the
'family' that overlook the nuanced relationships to which it might refer (Harker &
Martin, 2012; Wilkinson, 2014). With the exception of a few, the people in the
room were not related by blood, romantic attachment or domestic arrangement, yet
this description of a family implies the presence of interpersonal relations of care
and connection (Lawson, 2009). In this case, it is an intimacy that cuts across international borders, language, class, race, ethnicity, religion, gender, age, sexuality
and education and one that supports the claim that 'closeness' does not necessarily
map onto 'nearness' (Oswin & Olund, 2010; Cronin, 2015a). Just as the use of
family is noticeable for its intimated intimacy, the reference to 'relationships' as
the organization's core value is also noticeable. In prioritizing 'relationships' over
CLN's other values, such as justice, respect, dignity and equality, the building of
relationships is presented as both the foundation and goal for all of its work.

For Matt,[2] one of the European members, the relationships described are not
only 'authentic' but also central to CLN's running and motivation:

> We're all friends. I mean I think the thought is that we want to have authentic
> relationships. . . Over 15 years I have witnessed a couple of clashes [but] this
> is normal and I think if we hadn't had those strong authentic relationships,
> there would have been much more likelihood that they would break apart, for
> people to leave the organization, to turn their back. It's far from perfect and
> sometimes it is really tough business. But yeah, I think the fact that so many
> people really care for the other individuals in the room, is one way why [*sic*]
> we feel motivated to work through it.
>
> (Matt interview, USA, 2013)

As Matt suggests, friendships are not only an emotional mechanism of support
and care, but they work to maintain motivation through 'the tough business' of

social activism and volunteering, which can be undermined by personal clashes but also easily overshadowed by a variety of challenges and ever-diminishing support (Brown & Pickerill, 2009). At the time, these challenges included: severe cuts to funding and the resultant loss of paid positions and programme contracts; sharp rises in discrimination figures, including anti-immigrant sentiment and ongoing reports of police brutality; dwindling volunteer numbers; and a growing geographical inequity of resources. In this account, it is clear that while friend-ships and caring relationships have clearly emerged from the work, they are also crucial to its maintenance (Lawson, 2009).

It has long been recognized that face-to-face interaction, such as the kind ena-bled at the AGM, is crucial to developing and strengthening relationships and further generating the emotional energy that is so important to the sustainability of activism (Temenos, 2016). As Routledge et al. (2007, p. 2577) have argued, activist events act as 'zones of encounter' where ideas are shared, tactics are dis-cussed, resources are mobilized and links across different organizations, social movements, languages and localities are made (Halvorsen, 2015). These personal ties and energies provide the basis for collective identities, reduce isolation and enable the sharing of grievances and aspirations to create spaces where emotional reflexivity is possible (Brown & Pickerill, 2009). There are two points that are worth noting here. First, is that the AGM offered a chance to reflect on what was going well, which was clear in several of the key speeches:

> Kay reminds the audience of the 'joys of standing together' in the face of discrimination. She reinforces a sense of collective action and highlights the common position and values that hold the room together. Examples of achievement are presented, one after the other; numbers worked with; pro-jects started; awards received; training undertaken; new branches opened; and growing geographical reach. Whilst the figures speak for themselves, they are accompanied by statements and speeches by those that have worked with CLN, with each new announcement and report greeted with cheers, applause, whoops and the occasional standing ovation.
>
> (Observation, USA, 2014)

The emphasis on a common position and shared values is a common rhetoric of solidarity (Oosterlynck et al., 2016). Furthermore, this chance to celebrate and highlight achievement is crucial to the maintenance of motivation and provides a sense of collective purpose by bringing together the individual achievements from across the organization into one coherent narrative to be shared and celebrated together as a 'we'.

The second point to note is that while the energy generated from shared celebra-tion and joy serves to strengthen the sense of collective struggle, these events also function as 're-unions', in the sense that they become a space in which friendship and intimacies are reconfirmed and maintained. As Cronin (2015a, p. 668) has argued, while emotionally close friendships and intimacies can extend across con-siderable distances with only minimal contact, some form of co-presence is usually needed. In this regard, the less formal spaces of collaboration and knowledge exchange are also significant, demanding a move away from the formal spaces where speeches are delivered, to focus on the meeting rooms, dinner events, coffee and cigarette breaks, where information is shared, connections are made and agendas are set (Cook & Ward, 2012). This includes hotel breakfast buffets, downtown restaurants, mid-week daytrips, shared hotel rooms, elevators, the hotel gym and trips to the supermarket, all of which form a crucial part of the AGM week. These less formal spaces of socializing and enjoyment are spaces for further discussion and the generation of ideas but also where personal news and issues are discussed, emotions are analysed, support is extended, and relationships are nurtured and reproduced. Such informal and personal conversations inevitably 'frame' the more formal spaces of meeting (Ahmed, 2012).

In opening with an AGM I wanted to foreground the affective, 'familial' relationships that are at the core of CLN and emphasize how interpersonal relations are viewed in relation to CLN's wider activities. In so doing, it is possible to see how such meetings are about the renewal and maintenance of relationships as much as they are the celebration of social activism. With the remainder of the chapter I want to reflect on how these affective relations and intimacies are developed and sustained, to consider, as one volunteer described it, what got these individuals 'hooked' on this work in the first place.

4. Emotional work and sensuous solidarities

Many of the people in the room on that day in August 2014, were introduced to CLN through a leadership training event in their local area or workplace. These events see participants assemble in community centres, church halls or conference venues to undertake training around issues of prejudice and discrimination, violence or conflict. Some individuals were already involved in social activism or were 'diversity practitioners' (Ahmed, 2012) in one way or another and wanted extra coaching. Some attended for workplace training, while a few were following the recommendations of friends and family. A handful knew that they wanted to get involved in CLN or had signed up for volunteer work, while others were simply curious. Many were advocates of different causes – anti-racist agendas, youth work, indigenous rights, LGBTQ issues – and had their own aspirations and agendas, whether the development of personal leadership skills in their community, or addressing a particularly difficult form of discrimination that they had wanted to confront. What many people had in common was that they had not anticipated that these events would be the beginning of a much longer commitment and one that would see them working across differently situated struggles.

CLN workshops and programmes not only act as sites of encounter where people might learn about forms of discrimination experienced by others, but also where private struggles against prejudice and discrimination can be consciously crafted into more public claims for recognition and action. In short, they provide the ideal grounds for the development of solidarities or 'coalitions', which, as Nina suggested in the opening account, is one of CLN's main goals. In this section,

I want to focus on how the emotional and embodied experience of these events can become central to the building of coalitions across very different issues.

As I have suggested elsewhere (Wilson, 2013a), training workshops produce 'affective atmospheres' (Anderson, 2009). Tactics are used to shift the 'emotional tonality' (Conradson, 2003, p. 1986) of workshop spaces, enabling atmospheres to be 'performatively brought into being'. This includes ice-breaker sessions to ease tensions and break down barriers, and 'quiet time' to allow reflection and personal space after particularly difficult sessions or intense periods of translation. Clapping and dancing further enlivens workshop groups and regains attention, while music is variously used to relax, excite, calm and celebrate. Chairs are moved, seating arrangements are altered and participants are repeatedly expected to find new partners to work with so as to disrupt habits, test the limits of comfort zones and encourage participants to make connections with every other person in the room (Wilson, 2013a, 2014). The workshops involve hugging, crying, singing, laughing, the holding of hands and the telling of jokes.

Difficult and painful stories of discrimination are shared in small groups, with accounts told through tears, stammered words and hushed voices. These stories provide an opportunity to talk through experiences that have had an impact on people, and to further explore internalized oppression, but also offer other participants a chance to hear about the workings of discrimination that they may know little about. In the following example from a UK workshop, a man has just finished recounting his experience of a racist encounter. His personal account is used as a means to call for collective action:

> The workshop facilitator turns to the audience. 'You have just heard about the damaging and lifelong impacts of racism. I'm now looking for a show of hands – who here is willing to commit to better understanding the discrimination faced here – to being a better ally and to working with others to confront racism in whatever form it comes?' Hands go up. There is a heavy silence in the room. Some people are nodding firmly, others are fiddling with tissues and a couple of people are now standing with a look of defiance on their faces. The man at the front of the room is invited to take his time, look around and take note of all of the people that just heard his story and have agreed to do something with it. Minutes pass before he can bring himself to look up. When he does, the room is ready and hands are still raised.
>
> (Observation, UK, 2013)

The audience was diverse in terms of age, religion, class, ethnicity, nationality and sexual orientation and included charity, health and social workers, business managers, office workers, students and stay-at-home parents. While the man's account focused on an altercation with one man, the facilitator later reminded the audience that the focus should not be limited to the racist individual. As Ahmed (2012, p. 44) notes, a focus on 'bad individuals' understates the scope and scale of racism, allows the institutional and structural nature of racism to be ignored and leaves out a reflection on 'how racism *gets* reproduced'. Instead, the workshop facilitator connected this account of racism to persistent and wide-ranging structural inequalities, encouraging the audience to note the role of the justice system, educational inequality, immigration law, housing policy and the media. After hearing the story, the audience were asked to pledge a commitment to being a better 'ally' to the man by working with others to learn about, notice and confront racism in all of its varied forms.

Raising a hand in this instance is a performative act. While it might be argued that *acts* of solidarity might amount to nothing more than symbolic gestures, their significance, as Chatterton et al. (2013) argue, should not go unnoticed. This small and modest act, which is noticeably less spectacular than the 'acts' that tend to feature in academic research, does two things. First, it serves as a quiet comfort for those who have recounted their experience, serving to minimize the sense of isolation and acting as an acknowledgement that their circumstances have been heard and recognized by others in that moment. Second, the act of standing together or raising a hand in response to a personal story is intended to develop a collective sense of injustice and response. Participants are invited to acknowledge that their personal pledge is one made as part of a wider group – that they have been collectively *moved* to do so. Importantly, the commitment called for participants to collectively challenge social relations in what might be considered to be very quiet and banal ways, through 'noticing', learning about and contesting racism in their ordinary day-to-day lives. As a number of scholars have argued, this is the kind of action and context that should be at the heart of efforts to rethink how we understand social activism if we are to better value everyday action (Horton & Kraftl, 2009; Wilson, 2013a).

As Routledge (2012) has argued, the embodied and emotional experiences of political moments and events are also crucial to the building of solidarities (see also Askins, 2016). Shared bodily movements and engagements such as the rush of adrenalin, standing together, or the holding of hands, produce 'sensuous solidarities' (Routledge, 2012). These solidarities can arise from joy, anger or panic, sadness, disgust or pride. Transforming the 'emotional life' of participants is thus not only central to CLN's work, but develops an effective technique for action, a more productive group dynamic and a sense of common cause, while also aiming to effect 'personal transformation' (*CLN training material*).[3] In the example above, the audience is moved through a combination of shared anger and defiance as they respond to the man's story and raise their hands to show their collective support, demonstrating how outrage or anger might be cultivated to elicit a response and build an 'affective attachment to the cause' (Brown & Pickerill, 2009, p. 26). As one participant remarked in relation to a similar story:

> The thing is, it actually got me to the point of tears and everything else, especially when you find out (. . .) things about her life and see what she'd gone through, you know. So it was a very, very emotional. . . We both left the training room feeling like we'd done ten rounds with Mike Tyson or something. And that was it, I was hooked.
>
> (Eric interview, UK, 2014)

It is the emotional nature of the work that got Eric 'hooked'. This was how he described what was to be his first encounter with CLN and the start of his volunteer work with them. The experience he describes is embodied – likened to a physical fight – and is a typical narrative. As another volunteer stated: 'it just got me. . . in a way that nothing else ever had' (Simon interview, UK, 2014).

To this point, I have demonstrated how personal struggles resonate on an affective and emotional level to shape commitments to support differently situated others, but to also strengthen the desire to act and engage in volunteer work. However, while it would be easy to develop a linear account that would see emotions used as a tool for developing solidarity, I want to offer a more complicated narrative. Specifically, there is more to be said about how the exploration of emotions, the sharing of difficult experiences and the exposure to vulnerability, develops intimacies and interpersonal relations that work to create a less coherent picture of activism. As I will argue, it is this incoherence that is revealing of the complex ways in which solidarities are assembled and maintained, but also of how political action produces new forms of relation.

5. Building interpersonal relations and connecting diverse issues

> I shared things with the people in that room that I had never shared with anybody before – not even my family.
>
> (Harriet diary, UK, 2010)

This was a claim that came up repeatedly in my conversations with research participants. As Cronin (2015b) argues, normative views of intimate life regularly posit the family and (hetero)sexual couple relationships to be the most important in both practical and emotional terms. By suggesting that CLN events develop a space where participants share things that they hadn't '*even*' shared with their family, participants not only confirmed normative understandings of intimate life but in so doing, underlined just how significant – and surprising – the sharing of intimate details were.

Participants are not only encouraged to share their personal experiences of prejudice and internalized oppression, but to also openly explore their own prejudices. As Smith (2016) has suggested, the process of listening to somebody reveal intimate details about their life often results in an instinctual response – to offer up similarly intimate details. This reciprocity gives something back in a situation where others are willing to make themselves vulnerable. This is particularly key when trust and mutuality are central to the development of the kinds of solidarity CLN seeks to build (Routledge et al., 2007). Further still, sharing difficult experiences in this way, can work to create interpersonal relationships that evoke friendships (Rebughini, 2011; Smith, 2016). As sociological studies have highlighted, friendships can be forged and strengthened through the exploration of emotions and personal issues, and provide a space where it is easier to reveal vulnerabilities and seek help with the examination of 'your own life experience'

(Rebughini, 2011, p. 4.2). With CLN, I suggest that it is the rapidity with which these spaces are created that intensifies the felt attachment to others.

Exercises encourage participants to share information about their family and migratory histories, their race, religion, nationality and class. Sexuality, birth order, age and gender, along with hidden histories of illness, experience with the justice system, or connections with the armed forces, are also explored as key life experiences that shape one's sense of self. These exercises are not only a quick way of allowing participants to develop intimate knowledge about one another, but they also work to disrupt assumptions about sameness and difference, allowing participants to recognize the different ways in which they might connect with others in the room, whether through shared experiences of mental health, shared regional identity, a similar educational background, class-upbringing or religion. There are also opportunities for people to engage in 'culture shares' where they share something of personal significance with the rest of the group – an exercise that people are asked to prepare for in advance. The following example, taken from the US in 2014, is a typical account of a culture share that occurred at the end of a long day of workshop exercises.

> Films from family celebrations; traditional dress and fabrics; guitar music; black and white photographs of grandparents; poems written in school; short stories of home; migration histories; and pregnancy announcements, are just some of the accounts and objects shared by the group. Some people have brought biscuits along and there are coos over photographs, rounds of applause and small exclamations as items are passed around for a closer look. There are tearful smiles and extended arms. Then, after several hours of sharing, a dance that has the audience up on their feet and making the most of the floor space.
>
> (Observation, USA, 2014)

Opportunities to share stories and objects of personal significance are important. The passing around of materials and tactile engagement encourages interaction, while the music and shared dance movements give rise to collective feeling. Exchanged stories and tactile interactions such as those encountered here, are just as important to the development of reflective and reflexive memories (see also Askins & Pain, 2011), as the telling of difficult and intimate stories of prejudice and discrimination. By building relationships in these different ways, I suggest that we can see the development of what Askins (2015, p. 476) has called a 'quiet politics' of mutuality and care – that is, the development of relations that, while involving 'small acts' of care and concern, require an explicit 'political will to engagement' that demands a commitment. It is these bonds that intensify the desire to act and to build coalitions across different issues.

The bonds described above not only make it more likely that stories of prejudice and discrimination will continue to resonate beyond the initial training events, but they also mobilize the accounts of differently situated others. The friendships that I noted at the start of this chapter have, to borrow Conradson and

Latham's (2005, p. 291) words, 'remarkable spatial and temporal resilience' (see also Cronin, 2015a). They are not only sustained and actively 'nurtured' across large distances, through social media, emails or calls, but are further responsible for enacting patterns of mobility. In addition to annual meetings and training events, volunteers and members Facebook, Skype and attend retreats. They join wedding celebrations, birthdays and funerals. They enjoy shared dinners, visits and mini-breaks, and they exchange photos and emails about their families, children and day-to-day lives. These networks of friendship thus undergo 'various forms of stretching and distanciation' (Conradson & Latham, 2005, p. 295), which underline the 'agency of social networks in drawing people in certain directions, [and] in inviting and taking them to certain places' (ibid., p. 297). In short, despite the distance and the transnational nature of the network, the relationships developed go far beyond those that connect them to issues of social injustice. This level of intimacy across both difference and distance, and between people who, on the face of it, might have little in common, is exactly the kind of intimacy that Gandhi (2005) described as 'unlikely'. In the context of CLN, I suggest that it is also a form of intimacy that can be politically productive and one that highlights the excessive nature of voluntary spaces and how they 'leak out' into wider social practices, spaces and places (Smith et al., 2010).

When individuals travel, so too do stories of discrimination, exposing people to struggles and issues they had little knowledge of. For most people, their introduction to CLN was through a local leadership-training event. Because much of this training concerns the relationships between structural violence and individuals, the focus of the workshops tend to be shaped by those in the room, thus centring around local concerns or place-based struggles. In London, for example, race and religion are often two of the key issues discussed, whereas in the South West of England, class tends to feature more prominently as a topic for training and reflection. Beyond these workshops, and as individuals become more involved in CLN's work, they are introduced to a greater variety of issues and forms of advocacy as they interact with the wider organization and undertake training in different regions and countries.

As a researcher, I have moved from familiar discussions around race and religion in the context of a city that I know very well (Wilson, 2015), to not only thinking about race and religion in radically different local and national contexts, but engaging with work and issues in completely new areas. This has included indigenous rights in the US, work on sexism, sexual violence and homophobia, Latino discrimination, the stigma faced by atheists in the US South, along with a wide variety of place-based struggles, all of which are located in particular histories of action and inequality. Each of these new areas of concern have taken on greater significance as I have spent more time getting to know the people involved and witnessing the different ways in which they are personally affected. My research was full of examples of people who had become advocates for causes they had little previous knowledge of, all of which were connected to the caring relationships that they had developed with people in the organization. For example, this sees individuals who joined the charity with an interest in LGBTQ youth work, learning about and further supporting a variety of other campaigns, whether for Black African American struggles or distant place-based struggles such as those seen in Switzerland in 2009, when a ban to forbid the building of minarets was put to a public vote and passed (see Antonsich & Jones, 2010).

In contrast to the previous section, which focused on what emotions can do for social activism and solidarity – how, for instance, anger can become a tool for action – a focus on the role of emotional relationships, highlights how activist tendencies and solidarities are produced in unanticipated and unplanned ways (Horton & Kraftl, 2009). While I began the chapter with an account of spatial proximity, the distanciation of interpersonal relationships is key to the mobilization of issues and to connecting struggles and places. This is not always about strengthening mass action, but about developing understanding about differently situated others in ways that inflect everyday geographies elsewhere. By focusing on the divergent spaces that become entangled in CLN work and through which the ongoing commitment to work on inequality is ensured, the importance of questioning where activism gets done is underlined.

While I have emphasized the importance of care and friendship to the enactment of new mobilities, political action and learning, it is also important to recognize that 'unlikely' friendships demand work. In recognizing this, it would be remiss not to underline how political action also gives rise to – and further strengthens – friendship. The intimacies established in workshops are important to developing a will to act across a variety of causes, but political action is also significant to the maintenance of friendship and collective feeling. Indeed, what emerges is a complicated picture of mutually enforcing relationships, such that friendship and solidarity become impossible to separate in any meaningful way.

6. Conclusion

This chapter joins a growing number of contributions that have sought to highlight the personal and affective bonds that become important to social activism and forms of solidarity (Brown & Pickerill, 2009; Featherstone, 2012). In paying attention to the affective attachments that are developed across international borders, language, class, race, ethnicity, religion, gender, age, sexuality and education, the chapter has demonstrated how friendships arise from political action, sustain the motivation for social activism and also have the potential to further build coalitions across different struggles and place-based campaigns. As such, it underlines how friendships can be central to political change. Through attending to the *spaces* in which solidarities and friendships are built, the chapter not only challenges the assumptions of spatial boundedness that have shaped understandings of solidarity and its supposed links to communities of similitude, but has demonstrated the complex ways in which solidarities, which are both implicit and explicit in their intentions, develop. In seeking to move away from grand and linear narratives of struggle, to offer an account that foregrounds the multiple scales, gestures and often 'quiet politics' involved (Askins, 2015), the chapter highlights the often-overlooked and less spectacular ways in which social activism gets done, pointing to the deeply entangled and contingent nature of friendship and solidarity.

Importantly, the chapter does not claim that the connections, friendships and relationships discussed necessarily lead to successful campaigns, nor does it claim that all forms of solidarity necessarily involve friendship. While friendships are certainly valuable to the building of solidarities that seek to challenge inequality, not everybody has the ability to 'act politically across coalitions' (Routledge et al., 2007, p. 2576). The strength of personal relationships and solidarities can fluctuate with time and as the worries and commitments of everyday life take over, or other challenges present themselves, the urgency of these issues and their affective resonances can diminish. As was noted by Matt, the work undertaken by CLN is not easy. Gatherings and events are not without conflicts, contestations or inequalities (Wilkinson, 2009). There are competing agendas and political positions, differing language capabilities, knowledge and geographical imaginations, which make some individuals more or less capable of acting and allow some issues to resonate more than others (Routledge et al., 2007; Vasudevan, 2015). Furthermore, while the spaces of communication outlined here connect geographically discrete locations, they can also be shaped by the performance of reputational capital and geographical hierarchies (McCann, 2011), which can be exacerbated as resources fluctuate across the network.

With these caveats in mind, I want to underline what the chapter offers in terms of understanding what moves people to act. As Alderman et al. (2013) argue, this figurative account of movement is one that is rarely addressed when discussing the mobilization of resources for social change and action. Yet without it we would have little understanding of what sustains social action and voluntary practices (ibid.). As I have argued, attending to stories of friendship and emotional connection humanizes our understanding of advocacy and solidarity. In this case, it highlights that social activism is not always about political action that is intentional and linear, but about the unanticipated, messy and complex unfoldings of solidarities that are emotional, embodied and affective, but no less important. In focusing on what might be considered 'unlikely' friendships, which stretch across difference and distance, we see how people become compelled to support efforts for social change that previously had little personal resonance. We also see how friendships emerge out of activism – how solidarity and friendship are mutually built. In noting how these friendships are formed and strengthened through the sharing of intimate details and difficult experiences, it becomes clear that the strength of these relationships is central to sustaining motivation and securing commitment to voluntary work and campaigns across a number of years, the involvement in which further strengthens and affirms affective bonds.

In responding to the wider collection's interest in place, the chapter has also demonstrated how stories of discrimination are mobilized, to not only allow connections to be made across different struggles, but to allow a better understanding of how these struggles take shape in different local, national and international contexts. This includes a variety of place-based politics and struggles for recognition, the knowledge of which challenge and extend geographical imaginations. As

Routledge et al. (2007, p. 2759) have suggested, solidarities are not limited to virtual phenomena, but are rather 'physically manifest' in conferences, protests and meetings, and to this I would add homes and other, more intimate and less formal spaces. As such, the mobilities enacted are not only driven by the demands and requirements of social activist work, by also by leisure and friendship (Conradson & Latham, 2005; Cronin, 2015a). This mobility highlights the excessive nature of CLN sites and programmes, and demonstrates how struggles against inequality get caught up with other social practices and relationships that become impossible to separate.

The narratives documented in this chapter might not be considered remarkable, but they are fundamental to understanding how the desire to *do something*, or learn more, is cultivated, intensified and further sustained. As I have argued, these narratives underline the need for more 'enlivened' geographies of solidarity that focus on the banal and quiet ways in which differently situated others are brought together, and held together. In focusing on the transformative potential of friendship as an improvisational politics of co-belonging (Gandhi, 2005), it demonstrates how questions of intimacy and care can offer an alternative entry point into examinations of solidarity, to both complement and challenge existing accounts.

Acknowledgements

Enormous thanks to my research participants and friends at CLN who have made this work possible. My thanks also to the editors of the collection for detailed comments, to Saskia Warren for ongoing discussions and to Jonathan Darling for comments on earlier versions. This research was supported, in part, by the Royal Geographical Society (SRG 23.13), a British Academy/Leverhulme Small Grant (SG130284) and a R-SIF award from the University of Manchester.

Notes

1 CLN is a pseudonym.
2 All names are pseudonyms.
3 The details of the training material are anonymized to protect the organization's identity.

References

Ahmed, S. (2004). Collective feelings or, the impressions left by others. *Theory, Culture & Society, 21*, 25–42.
Ahmed, S. (2010). *The promise of happiness*. Durham, NC: Duke University Press.
Ahmed, S. (2012). *On being included: Racism and diversity in institutional life*: Durham, NC: Duke University Press.
Alderman, D. H., Kingsbury, P., & Dwyer, O. J. (2013). Reexamining the Montgomery bus boycott: Toward an empathetic pedagogy of the Civil Rights Movement. *The Professional Geographer, 65*, 171–186.
Anderson, B. (2009). Affective atmospheres. *Emotion, Space and Society, 2*, 77–81.

Antonsich, M., & Jones, P. I. (2010). Mapping the Swiss referendum on the minaret ban. *Political Geography, 2*, 57–62.

Askins, K. (2015). Being together: Everyday geographies and the quiet politics of belonging. *ACME: An International E-Journal for Critical Geographies, 14*, 470–478.

Askins, K. (2016). Emotional citizenry: Everyday geographies of befriending, belonging and intercultural encounter. *Transactions of the Institute of British Geographers*, doi: 10.1111/tran.12135.

Askins, K., & Pain, R. (2011). Contact zones: Participation, materiality, and the messiness of interaction. *Environment and Planning D: Society and Space, 29*, 803–821.

Berlant, L. G. (2000). *Intimacy*. Chicago, IL: University of Chicago Press.

Bowlby, S. (2011). Friendship, co-presence and care: Neglected spaces. *Social & Cultural Geography, 12*, 605–622.

Brown, G., & Pickerill, J. (2009). Space for emotion in the spaces of activism. *Emotion, Space and Society, 2*, 24–35.

Bunnell, T., Yea, S., Peake, L., Skelton, T., & Smith, M. (2012). Geographies of friendships. *Progress in Human Geography, 36*, 490–507.

Chatterton, P., Featherstone D., & Routledge, P. (2013). Articulating climate justice in Copenhagen: Antagonism, the commons, and solidarity. *Antipode, 45*, 602–620.

Conradson, D. (2003). Spaces of care in the city: The place of a community drop-in centre. *Social & Cultural Geography, 4*, 507–525.

Conradson, D., & Latham, A. (2005). Friendship, networks and transnationality in a world city: Antipodean transmigrants in London. *Journal of Ethnic and Migration Studies, 31*, 287–305.

Cook, I. R., & Ward, K. (2012). Conferences, informational infrastructures and mobile policies: The process of getting Sweden 'BID ready'. *European Urban and Regional Studies, 19*, 137–152.

Cronin, A. M. (2014). Between friends: Making emotions intersubjectively. *Emotion, Space and Society, 10*, 71–78.

Cronin, A. M. (2015a). Distant friends, mobility and sensed intimacy. *Mobilities, 10*, 667–685.

Cronin, A. M. (2015b). Gendering friendship: Couple culture, heteronormativity and the production of gender. *Sociology, 49*, 1167–1182.

Darling, J. (2011). Giving space: Care, generosity and belonging in a UK asylum drop-in centre. *Geoforum, 42*, 408–417.

Darling, J., & Wilson, H. F. (2016). *Encountering the city: Urban Encounters from Accra to New York*. London: Routledge.

Derrida, J. (2005). *The Politics of friendship*. London: Verso.

Desai, A., & Killick, E. (2013). *The ways of friendship: Anthropological perspectives*. New York: Berghahn Books.

Farías, M. (2016). Working across class difference in popular assemblies in Buenos Aires. In J. Darling & H. F. Wilson (Eds.), *Encountering the city: Urban encounters from Accra to New York*. London: Routledge, 169.

Featherstone, D. (2010). Contested relationalities of political activism: The democratic spatial practices of the London Corresponding Society. *Cultural Dynamics, 22*(2), 87–104.

Featherstone, D. (2012). *Solidarity: Hidden histories and geographies of internationalism*. London: Zed.

Foucault, M. (1997). Friendship as a way of life. In P. Rabinow (Ed.), *Ethics, subjectivity and truth: The essential works of Michel Foucault 1954–1984*. New York: The New Press.

Fyfe, N. R., & Milligan, C. (2003). Out of the shadows: Exploring contemporary geographies of voluntarism. *Progress in Human Geography, 27*, 397–413.

Gandhi, L. (2005). *Affective communities: Anticolonial thought, fin-de-siècle radicalism, and the politics of friendship*. Durham, NC: Duke University Press.

Halberstam, J. (2005). *In a queer time and place: Transgender bodies, subcultural lives*. New York: NYU Press.

Hall, S. M., & Jayne, M. (2016). Make, mend and befriend: Geographies of austerity, crafting and friendship in contemporary cultures of dressmaking in the UK. *Gender, Place & Culture, 23*, 216–234.

Halvorsen, S. (2015). Encountering Occupy London: Boundary making and the territoriality of urban activism. *Environment and Planning D: Society and Space, 33*, 314–330.

Hardt, M., & Negri, A. (2001). *Empire*. Cambridge, MA: Harvard University Press.

Harker, C., & Martin, L. (2012). Familial relations: Spaces, subjects, and politics (guest editorial). *Environment and Planning A, 44*, 768–775.

Horton, J., & Kraftl, P. (2009). Small acts, kind words and 'not too much fuss': Implicit activisms. *Emotion, Space and Society, 2*, 14–23.

Ince, A. (2015). From middle ground to common ground: Self-management and spaces of encounter in organic farming networks. *Annals of the Association of American Geographers, 105*, 824–840.

Kathiravelu, L. (2013). *Friendship and the urban encounter: Towards a research agenda* (MMG Working Paper 13-10). Retrieved from Max Planck Institute for the Study of Religious and Ethnic Diversity at http://pubman.mpdl.mpg.de/pubman/item/escidoc: 1850592/component/escidoc:1850591/ WP_13-10_Kathiravelu_Friendship.pdf.

Lawson, V. (2009). Instead of radical geography, how about caring geography? *Antipode, 41*, 210–213.

Massey, D. (2005). *For space*. London: SAGE.

McCann, E. (2011). Urban policy mobilities and global circuits of knowledge: Toward a research agenda. *Annals of the Association of American Geographers, 101*, 107–130.

McFarlane, C. (2006). Transnational development networks: Bringing development and postcolonial approaches into dialogue. *The Geographical Journal, 172*, 35–49.

McFarlane, C. (2009). Translocal assemblages: Space, power and social movements. *Geoforum, 40*, 561–567.

Mills, S. (2013). Surprise! Public historical geographies, user engagement and voluntarism. *Area, 45*, 16–22.

Neal, S., & Vincent, C. (2013). Multiculture, middle class competencies and friendship practices in super-diverse geographies. *Social & Cultural Geography, 14*, 909–929.

Oosterlynck, S., Loopmans, M., Schuermans, N., Vandenabeele, J., & Zemni, S. (2016). Putting flesh to the bone: Looking for solidarity in diversity, here and now. *Ethnic and Racial Studies, 39*, 764–782.

Oswin, N., & Olund, E. (2010). Governing intimacy. *Environment and Planning D: Society and Space, 28*, 60–67.

Pain, R., & Staeheli, L. (2014). Introduction: Intimacy-geopolitics and violence. *Area, 46*, 344–347.

Povinelli, E. A. (2006). *The empire of love: Toward a theory of intimacy, genealogy, and carnality*. Durham, NC: Duke University Press.

Pratt, G. (2008). International accompaniment and witnessing state violence in the Philippines. *Antipode, 40*, 751–779.

Rebughini, P. (2011). Friendship dynamics between emotions and trials. *Sociological Research Online, 16*, 3.

Routledge, P. (2012). Sensuous solidarities: Emotion, politics and performance in the Clandestine Insurgent Rebel Clown Army. *Antipode, 44*, 428–452.

Routledge, P., Cumbers, A., & Nativel, C. (2007). Grassrooting network imaginaries: Relationality, power, and mutual solidarity in global justice networks. *Environment and Planning A, 39,* 25–75.

Schuermans, N. (2013). Ambivalent geographies of encounter inside and around the fortified homes of middle class Whites in Cape Town. *Journal of Housing and the Built Environment, 28,* 679–688.

Smart, C., Davies, K., Heaphy, B., & Mason, J. (2012). Difficult friendships and ontological insecurity. *The Sociological Review, 60,* 91–109.

Smith, F. M., Timbrell, H., Woolvin, M., Muirhead, S., & Fyfe, N. (2010). Enlivened geographies of volunteering: Situated, embodied and emotional practices of voluntary action. *Scottish Geographical Journal, 126,* 258–274.

Smith, S. (2016). Intimacy and angst in the field. *Gender, Place & Culture, 23,* 134–146.

Temenos, C. (2016). Mobilizing drug policy activism: Conferences, convergence spaces and ephemeral fixtures in social movement mobilization. *Space and Polity, 20,* 124–141.

Valentine, G. (2008). Living with difference: Reflections on geographies of encounter. *Progress in Human Geography, 32,* 323–337.

Vasudevan, A. (2015). *Metropolitan preoccupations: The spatial politics of squatting in Berlin.* Oxford: Wiley-Blackwell.

Vertovec, S. (2015). *Diversities old and new: Migration and socio-spatial patterns in New York, Singapore and Johannesburg.* Basingstoke: Palgrave Macmillan.

Warren, S. (2014). 'I want this place to thrive': Volunteering, co-production and creative labour. *Area, 46,* 278–284.

Wilkinson, E. (2009). The emotions least relevant to politics? Queering autonomous activism. *Emotion, Space and Society, 2,* 36–43.

Wilkinson, E. (2014). Single people's geographies of home: Intimacy and friendship beyond 'the family'. *Environment and Planning A, 46,* 2452–2468.

Wilson, H. F. (2012). Living with difference and the conditions for dialogue. *Dialogues in Human Geography, 2,* 225–227.

Wilson, H. F. (2013a). Collective life: Parents, playground encounters and the multicultural city. *Social & Cultural Geography, 14,* 625–648.

Wilson, H. F. (2013b). Learning to think differently: Diversity training and the 'good encounter'. *Geoforum, 45,* 73–82.

Wilson, H. F. (2014). The possibilities of tolerance: Intercultural dialogue in a multicultural Europe. *Environment and Planning D, 32,* 852–868.

Wilson, H. F. (2015). An urban laboratory for the multicultural nation? *Ethnicities, 15,* 586–604.

Wilson, H. F. (2016). On geography and encounter: Bodies, borders, and difference. *Progress in Human Geography,* doi: 10.1177/0309132516645958.

5 Self-building in Northern Italy

Housing and place-based solidarities among strangers

Michela Semprebon and Martina Valsesia

Introduction

Housing policies in Italy have long favored ownership, thus producing a dual housing system characterized by a small rental sector. Shrinking public investment and sales to sitting tenants have led to insufficient social housing supply (less than 4% of the total housing stock in 2011, against a EU average of 18%) (Scanlon et al., 2014). Housing prices have increased by 100 per cent from 1998 to 2007 and rents by 70 per cent (Banca d'Italia, 2013). The owner-occupied sector has experienced a reduced compliance in the coverage of mortgage debts, while the private rental market has seen an increase in prices (Baldini & Poggio, 2014), together with a multiplication of evictions by two and a half times compared to ten years ago (CGIL, 2013).

For immigrants, access to affordable housing is certainly a problem. Even more so since the economic crisis, immigrant families have been facing difficulties in meeting rent and mortgage costs (Alietti & Agustoni, 2013). Immigrants have been further penalized by enduring discrimination in the housing market and, in the Lombardy region, by a differential access to social housing (conditional upon a residence permit or a regular job contract in the regional territory).

Against this background, the revival of self-building has come along as a partial solution to facilitate immigrants' access to affordable housing. However, self-building has been mostly analyzed from an architectural perspective, with attention to technical and project management processes (see Bertoni & Cantini, 2008 for a review). Hardly any attention has been paid to its sociological dimensions. This chapter provides a contribution in this direction, by focusing on two self-building projects in the Lombardy region, in the north of Italy. In a scenario characterized by shrinking public investment in social housing, enduring forms of discrimination towards migrants and the multiplication of evictions, we will investigate how these self-building projects have paved the way to various forms of solidarity. We will explore how engagement in the actual process of self-building has nurtured solidarities, how these solidarities have supported the implementation of the projects themselves and how solidaristic bonds have been challenged throughout the projects.

First, we will provide a brief theoretical overview of the literature on social mix and community development that is relevant to this chapter. Second, we will sketch out the historical background of self-building in Italy and its institutional setting. Then, we will move to the empirical sections. Finally, we will draw some conclusions.

Social mix and community development

In the field of housing studies, considerable attention has been paid to social mix projects and their potentialities and limits to generate solidarity among strangers. Such projects have been largely based on the assumption that proximity among different groups can stimulate encounters and interaction which, in turn, can nurture solidarity (Kearns & Forrest, 2000; Dekker & Bolt, 2005). It is widely thought that interaction between strangers can be enhanced by mixing them at the building block level (Atkinson & Kintrea, 1998; Kleit, 2008; Van Kempen & Bolt, 2009).

Discussions about social mix in housing developments relate to much wider issues about the nature and the purposes of interactions in urban areas. As Simmel (1950) stressed, even apparently trivial contacts can hold society together. Attachment among strangers can be favored by proximity, especially if people engage in activities that stimulate a shared learning process (Dewey, 1954; Amin, 2002). Nevertheless, co-presence with diverse people is not sufficient per se to stimulate substantial interaction (Loopmans, 2000), nor the emergence of solidarity. On the one hand, living together may involve tensions, especially at a micro-scale (Goodchild & Cole, 2001; Kleinhans et al., 2007); on the other, it can result in mere incidental encounters (Valentine, 2008; Matejskova & Leitner, 2011) or in individuals leading separate lives (Atkinson & Kintrea, 1998; Allen et al., 2005; Arthurson, 2012; Graham et al., 2009).

Researchers have underlined the relevance of community development/building actions to promote residents' participation and increase contact in mixed housing projects (Silverman et al., 2006; Camina & Wood, 2009; Chaskin & Joseph, 2010; Mugnano & Palvarini, 2013.). By way of example, Joseph (2006) stresses the importance of informal events in facilitating interpersonal connections and in making common needs and interests emerge. Silverman et al. (2006) suggest informal events encourage social contact among inhabitants. Likewise, Camina and Wood (2009) argue that community development can positively impact on dwellers' engagement to get activities going. Importantly, the debate on community development activities has led to growing attention for the multi-dimensional character of housing management in the housing sector. CECHODAS – the European Committee for the promotion of the right to housing – has called for an integrated management that takes into consideration both the real estate and the social aspect of housing, thus encouraging the planning and implementation of programs that facilitate cohabitation among neighbors. The explicit objective to respond to individual housing needs is linked to the needs of the wider community, hence their interest in community development/building activities.

In Italy, the implementation of integrated housing management has included the introduction, in some social housing projects, of the '*gestore sociale*' (Ferri, 2011). This role builds on the centenary tradition of housing cooperatives and on the self-management of spaces and services within subsidized public housing (Galeazzi & Valsesia, 2013). The *gestore sociale* has responsibility over property and facility management as well as community management. The specific task is different from ordinary real estate managers and consists in engaging with dwellers and their needs by operating proactively in the organization of community life and in the planning of collaborative services. Hence, the *gestore sociale* has the task of preventing and managing conflicts, facilitating socialization and activating resources and skills that can in turn mould solidarity.

In the specific case of self-building, the so-called '*mediator*' is intended to raise self-builders' awareness of their skills and capacities to engage in the process. The *mediator* is also required to stimulate socialization and to prevent conflicts slowing down the construction process and impeding the emergence of solidarity bonds among self-builders. In spite of a limited debate in the academic literature, Golinelli (2006) argues, in line with the view of design professionals (Vestbro, 2000; Meroni & Sangiorgi, 2011), that mediators can have a positive impact, in the wider sphere of facilitation/participatory approaches, by promoting positive cohabitation and inclusion.

From this perspective, the self-building projects presented here can be understood as shared housing practices in which self-builders appropriate and learn a construction technique and the relevant skills in order to build their own house and live in it. They can be described as everyday place-based practices shaped by continuous negotiations (Lepofsky & Fraser, 2003). In line with Lefebreve's conceptualization of 'inhabiting a house' (1974), they can be seen as projects involving the appropriation of a space, rather than its mere ownership. Bearing this in mind, we will analyse the effects of two self-building projects and highlight how they promoted various forms of solidarity, while also digging out the ambivalent aspects that challenged solidaristic drives. Drawing on Oosterlynck et al. (2016), we will investigate the sources of these forms of solidarity. As elaborated in the introduction to this volume, this can entail encounters, shared norms and values, interdependencies that emerge from the division of labour and struggles around a common interest. In our analysis, particular attention will be devoted to the role of the 'mediator' in encouraging and strengthening social contacts and positive social relations.

Self-building

Self-building has long been a part of humans' housing practices (Paolella, 2008, p. 14). Dwellers have long been active actors in the construction, management and change of their dwellings, both in collective projects involving the spontaneous construction of entire villages (Fundarò, 1977) and in projects aimed at the construction of family units. It was only in modern times that many dwellers were excluded from the building process (Paolella, 2008). In the context of a rigid normative and bureaucratic apparatus, the construction of dwellings was devolved more and more to specialized professionals. This is how dwellers, as actors, have gradually transformed into consumers. In the words of Illich (1974, p. 146), 'the right to act has been replaced by the right to have access to'. In any case, self-building could well contribute to bringing back the agency of dwellers (Ceragioli & Martiano, 1985).

Recent academic literature on the topic is rather scarce (for an overview, see Harris, 1999; Colombo et al., 2010; Benson, 2015). With reference to developing countries, some authors argue that self-building is a consequence of a regulatory failure which holds back 'formal' housing markets (e.g. Ball, 2006). In advanced economies, it is often associated with affluent individuals willing to bypass restrictive labor market legislation, taxation, building and planning controls to get a dwelling perfectly suited to their needs. The practice of 'assisted self-building', on which this chapter concentrates, is a distinctive type of housing supply. This form of self-building can be described as:

> a particular building procedure, characterised by specific and consolidated construction methods and technologies, managed and co-ordinated by professionals, through which an associated and voluntary group of people and/ or families build, in their free time and/or during non-working hours, their own house.
>
> (Colombo et al., 2010, p. 93)

The idea that institutions could help families in the construction of their own house was first brought forward by Richard S. Harris, in Sweden, in 1904. The expression 'aided self-help housing' was already coined in 1945 by Jacob L. Crane, who theorized the concept and put it into practice in Peru. Since the 1970s, various forms of self-building have been established in Northern Europe, the Mediterranean, North and Latin America, Asia and Australia (see Bertoni & Cantini, 2008).

In Italy, self-building emerged from a strong class consciousness that led to the development of building cooperatives and self-build movements, through the pooling together of different resources and skills from family members and friends. The phenomenon was associated with a pre-capitalistic rural type of economy (Fera & Ginatempo, 1985): a family unit managed (directly) all the phases of the building process, to address the material needs for dwellings and to reduce costs. In the 1970s, a boom of self-building projects was recorded.

In the 1990s, the first forms of associated and assisted self-building spread in the Lombardy region, thanks to the organizational and promotional skills of architect Cusatelli. Cooperatives participated in public tenders for access to public land (Colombo et al., 2010) and carried out the projects directed by the same architect. In 2000, associated and assisted self-building gained momentum. Differently from the past, this new form involves middle-income individuals both of Italian and foreign nationality. While these people experience difficulties in accessing affordable housing, they are not among the most vulnerable in society. Additionally, the self-builders engaging in associated and assisted self-building do not aggregate voluntarily on the basis of family or friendship bonds, but are selected by an organization.

Introducing the case studies[1]

According to Colombo et al. (2010), 35 assisted self-building projects could be counted in Italy in the period 2000–2011. Many of them were set up by Alisei, a NGO with long-term experience in international cooperation and self-building in developing countries. Following the initiation of a few participatory projects in the Umbria region, media attention to self-building started growing. Consistent effort was made to promote it. Soon, Alisei became the main self-building referent of local (and regional) government agencies.

In Lombardy, a Regional Experimental Program on self-building (REP) was approved in 2005. It was supposed to last for three years. It was one of the largest schemes in Europe, aiming at building 650 flats. The Regional Authority co-financed 20 per cent of it (11.8 million euro); the remaining 80 per cent would be paid by the self-builders themselves. Through a convention with a private building society or a regional social housing agency, each project was promoted by the local authority in which the project was located. Only few local authorities joined the program, however, due to the limited time for application and/or limited available land for self-building. The REP eventually led to six projects, each one including 10 to 20 housing units. None of the projects was completed, except for the one in Paderno Dugnano.

Generally, four main actors can be identified in the process of self-building (see Table 5.1). First, a local authority identifies and provides an area for self-building, appoints a building company to manage the project, publishes a public tender to promote the project, invites interested self-builders and supervises the project. Second, a building society manages the project, draws up the financial plan, selects the self-builders based on objective and subjective criteria,[2] manages and coordinates the project and eventually assigns the flats to each family unit, by means of a raffle. The building society includes one or more site managers/architects, who provide technical expertise, and a mediator, who works in close collaboration with the architects. Third, a self-building cooperative provides the labour (normally for at least 60 hours/per month over a period of at least two years – although all the projects of the REP extended over four to five years). Fourth, a financial partner provides the mortgage. Self-builders are asked to pay a rent for ten years and a final amount to redeem the property.

The completed REP project in Paderno Dugnano is our first case study. It is located in a medium-sized town in the metropolitan area of Milan with approximately 47,000 inhabitants, of whom 7.2 per cent are of immigrant origins (more than the 4.7% in 2006, but still lower than the regional average of 10%, according to 2011 census data). The self-building project comprises a block of ten apartments. Work on the building site started in January 2006 and the apartments were formally completed in October 2012. The project involved a balanced mix of 10 families of Italian and immigrant origins (from Tunisia, Colombia and Albania). In the first year, three families dropped out and were replaced by three others from a reserve list. Table 5.1 provides a list of the main stakeholders.

Table 5.1 The actors of the self-building process

Actor	Paderno dugnano	Casalmaggiore
LOCAL AUTHORITY	Municipality of Paderno	Municipality of Casalmaggiore
BUILDING SOCIETY	Alisei NGO and then Alisei Autocostruzioni srl (operative arm of Alisei NGO)	Alisei NGO and then Alisei Autocostruzioni srl (operative arm of Alisei NGO) ***following Alisei bankruptcy, the project was managed by the social enterprise 'Sottoiltetto'
SELF-BUILDING COOPERATIVE	Cooperativa Progetto Uniti	Cooperativa Un tetto per tutti
FINANCIAL PARTNER	(Non-profit) Banca Etica	(Non-profit) Banca Prossima

The second case study is a project implemented in Casalmaggiore, a small town in the south of Lombardy, with approximately 15,000 inhabitants. In Casalmaggiore, the percentage of inhabitants with migrant origins is higher than in Paderno Dugnano, namely 10 per cent in 2006 and 15 per cent in 2011. The self-building project comprises three blocks of 19 apartments and did not receive REP funding. Work on the building site started in April 2007 and was completed in 2011. The project involved 18 families of Italian and immigrant origins (from Ghana, Albania, Morocco and Algeria). Various families withdrew from the project that eventually counted 12 families. Again, Table 5.1 provides a list of the main stakeholders.

Our analysis draws from a total of 21 semi-structured interviews: 13 with self-builders (two in Paderno Dugnano,[3] 11 in Casalmaggiore); three with representatives of Alisei building society (the President and two architects that managed and supervised the self-builders' team in Paderno Dugnano); one with a project leader of the Regional Experimental Program on Self-Building (REP); one with a public officer of the Housing Department of the Paderno Dugnano Municipality; one with an officer of ALER Regional Social Housing Agency; and two with representatives of the Cooperativa Sottoiltetto in Casalmaggiore (the mediator and the financial/project manager).[4]

In the next two sections, we will present the results of the analysis. First, we will focus on the development of forms of solidarity among the self-builders and on the sources of this solidarity. The subsequent section will point to the limits of solidarity among self-builders.

Hic et nunc practices of solidarity

Respondents generally described the self-building process as a largely positive learning trajectory that grew them stronger as a community (see Dewey, 1954). The first year of construction works were marked by full enthusiasm. Morale was high and relationships of mutual trust developed. Everyday work at the building site, the sharing of labor and fatigue formed the basis of strong feelings of community:

> At the beginning it was all very exciting! We all worked a lot, some of us worked every day and some were even asked to stop a little as they were working too much.
>
> (Self-builder 1, Venezuelan, Paderno)

> Today, again, we have worked together, we have discussed (. . .). So it will not be so difficult to live together. I do not think there will be any problem. We have grown very united [as a community].
>
> (Self-builder 1, Ghanaian, Casalmaggiore)

Interaction among individuals and family units was fostered by on-site work, but also by repeated opportunities for socialization, such as parties organized in celebration of important steps and achievements in the construction process:

> We organized dinners, we ate together. And also when we [accidentally] met in the town square we stopped to have a drink together. I feel very strong friendships have developed.
>
> (Self-builder 3, Albanian, Casalmaggiore)

In the case of Paderno, the availability of common spaces was also reported, by one of the architects, as important in facilitating communal activities:

> The first year it all ran smoothly, there was a very good balance within the group, with considerable involvement by all self-builders and big parties and dinners organized by their families. (. . .) It helped to have some common spaces.
>
> (Architect, Italian, Paderno)

Both the municipal officer and the self-builders stressed the role of the mediator to support the self-builders in their common endeavour:

> The mediator was good. She managed to keep people together. She organized meetings to encourage familiarization. Yes, the social goal of the project was clear to me.
>
> (Municipal officer, Italian, Paderno)

> I soon became very close to the mediator. She worked well to manage conflicts. She listened to us a lot. She did help.
>
> (Self-builder 1, Venezuelan, Paderno)

It is evident how the practice of self-building built on the encounter of self-builders and the sharing of their experience. It emerged as a 'relational form of inhabiting', as opposed to a 'solitude without isolation' (Augé, 2002). Self-building acted to prevent the anonymity, the solitude and the limited opportunities for sharing material and non-material resources that are often associated with life in a building block:

> We are not a group of strangers, as if we had to buy a house (. . .) [with] ten neighbors that look at us from far as they do not know us. (. . .) We all share something; we greet each other and I think that when the construction works will be over we will share even our houses (. . .) I do not feel alone and I do not think I will ever feel alone here.
>
> (Self-builder 2, Italian, Casalmaggiore)

Similar feelings were expressed also by immigrant self-builders, with particular reference to their inclusion:

> I have been living here for more than ten years (. . .). If an Italian does not know you, he/she does not talk to you. He/she sees you and that's it. Here, we all greet each other and talk to each other.
>
> (Self-builder 9, Moroccan, Casalmaggiore)

What emerges from the above extract is that the 'Other' was not feared, nor rejected or excluded. On the contrary, the Other was very much valued and active efforts were made to support inclusion in the community:

> I have learnt to relate to Muslims, Christians, Italians, Africans, many different people. (. . .) After a few months we were together as a family.
>
> (Self-builder 9, Moroccan, Casalmaggiore)

In particular, immigrant respondents conveyed how the project was important in this sense. They described the actual act of self-building as a turning point for them not to feel strangers any longer:

> I hope I will continue living here, as I do feel home, I am building this house, I do not feel I am a stranger anymore.
>
> (Self-builder 1, Ghanaian, Casalmaggiore)

> Thanks to the house we have built, I feel more attached to this town (. . .). This place is now becoming my home.
>
> (Self-builder 9, Moroccan, Casalmaggiore)

Well beyond the generation of bonds between family units, the intense experience of self-building also strengthened bonds within the families themselves, thus promoting positive intergenerational interactions:

> Thanks to this project I have found out many things about myself, my husband, my family. (. . .) My family has been growing stronger.
>
> (Self-builder 8, Moroccan, Casalmaggiore)

Through the self-building process, feelings of attachment developed strongly in relationship to the actual place (see Cresswell, 2004; Harvey, 2009). Self-building is evidently an example of place-based practice characterized by proximity and intimacy (see Jessop et al., 2008), whereby an experiential level of interaction leads to an emotional one.

Referring to the sources of solidarity, the relationship among the self-builders grew stronger through the interdependencies associated with the construction works. A good division of heterogeneous and complementary tasks and skills was required for their common goal to be achieved. They needed to help each other and to contribute to the progress of the project. This had a positive impact on self-builders' engagement:

> One's own specific engagement is no longer the engagement of one individual only that draws help from the group, but becomes part of the group's engagement (. . .) Thank God we also have a bricklayer in the group. And so he took up this task and helped to coordinate the group a little as well.
>
> (Self-builder 2, Italian, Casalmaggiore)

> All of us had a bit of experience, not necessarily professional experience, in the building sector: one works in public transport, a few are salesmen, etc. and this helped the whole group.
>
> (Self-builder 1, Venezuelan, Paderno)

> I have to say that I also taught some of the self-builders how to handle the tools as they were not very used to them.
>
> (Self-builder 3, Albanian, Casalmaggiore)

While engagement in the common construction activity favored attachments among self-builders and got the building work going, it was increasing awareness of each other's interdependency that allowed solidarity to develop. It was interdependency not in a Marxist sense: it developed across class lines. Yet, the common interest and shared work alone would have not been sufficient to create a stable basis for solidarity (see Crow, 2002). Encounter was a crucial element that contributed to the growth of solidarity beyond family units. Self-building fostered more than fleeting encounters. It generated informal relationships that developed alongside principles of reciprocity, self-help and collaboration and towards the re-elaboration of a sort of 'communal living':

> Many of [the children] were very small. (. . .) It often happened that while men [self-builders] were at the construction site, women would go to the park with the children and prepare lunch for everyone. (. . .) This is how it worked. People got together and everybody helped each other. And solidarity is still there.
>
> (Self-builder 1, Venezuelan, Paderno)

> When a difficulty arises, we help each other. (. . .) We do not know what the future holds, but if everything continues like this, we will keep helping each other.
>
> (Self-builder 11, Ghanaian, Casalmaggiore)

I have thought that my son will play with other kids here and one of us will be taking care of them in turn. And I very much like this.

(Self-builder 9, Moroccan, Casalmaggiore)

These extracts project solidarity generated through practices in the 'here and now' to the 'there and then'. The horizon of these solidarities lies beyond the end of the construction project. For example, self-builders referred to the fact that family units could collaborate with each other in taking care of the children, thus taking the burden away from family members engaged in self-building, but also providing the children with a welcoming supportive environment, so that they would also get to know each other. However, apart from one, no self-builder projected the future scenario of grown up children moving out of the newly built dwellings. The question is, thus, whether bonds of solidarity can evolve over time, throughout the life cycles of families and the various phases that characterize them.

Individualistic attitudes and conflictual dynamics

The forms of solidarity described in the previous section were very much challenged throughout the projects. Two concomitant factors contributed to changing the overall positive picture described so far: individualistic attitudes and conflictual dynamics. Together with some limits of the actual projects, such as the partial effectiveness of the mediator and the lack of supervision on the side of the local authorities, these factors interfered in the smooth progress of construction works.

In Casalmaggiore, self-builders reported on misunderstandings that surfaced already in the very first phase of the construction works. On the one hand, people struggled to 'read' the architectural project, to find out later that the actual houses were very different from their expectations. On the other hand, they also did not fully understand the actual requirements for their engagement in the project. This was particularly the case among families of immigrant origins – due, among other things, to limited fluency in the Italian language – and among families that joined the project at a later stage to replace members that had dropped out, as they had missed part of the initial information sessions.

We did not understand what it was like (. . .) They told us: we have explained all of it for two weeks! (. . .) But how can you understand by looking at a piece of paper? If you want to buy a house, you go there and walk inside it.

(Self-builder 1, Ghanaian, Casalmaggiore)

At the beginning, I got it wrong: I thought we would help, because we did not have any experience of building. It was not clear for me that we actually had to do all the work.

(Self-builder 1, Ghanaian, Casalmaggiore)

Difficulties continued to emerge as the construction works progressed. This fuelled conflicts and resulted in a vicious circle of misunderstandings among self-builders and between the latter, the architects and operators of the building society, especially in Casalmaggiore:

> Look, to tell you the truth, we missed some information. (. . .) They did not make themselves understood. They did not explain the message properly.
>
> (Self-builder 8, Moroccan, Casalmaggiore)

Against this background, self-builders lost enthusiasm and motivation started fading away. This is also why some of them eventually decided to abandon the project:

> We lost hope, we had no motivation to continue, we were scared! And this is terrible for this kind of project. (. . .) If you lose hope, the works slow down, everything gets stuck. (. . .) And this is how many people decided to quit.
>
> (Self-builder 8, Moroccan, Casalmaggiore)

According to one of the architects in Paderno, difference in norms and values, and in particular a different perception of what is a 'dwelling', may have impacted negatively on the enthusiasm of self-builders and their feelings of belonging. As he explained:

> Italians manifested feelings of emotional attachment, with reference to a 'home' for their family. Immigrants, on the contrary, described their house as a mere 'shelter'.
>
> (Architect, Italian, Paderno)

A second issue that affected the construction process negatively relates to inter-dependency as a source of solidarity. As anticipated, when construction works started, self-builders developed increasing awareness of their mutual forms of obligation. People got to know each other and found an equilibrium in the division of labor in the project. This contributed strongly to the generation of solidarity. This solidarity was put to test, however, when people started to leave the project and when others failed to take up the tasks that had been assigned to them. In general, construction works are organized in groups of two people. If one is missing or does not do what he or she is supposed to do, the other cannot add much to the labor force either. Accordingly, absenteeism was a real problem. Project leaders repeatedly insisted on the importance to engage punctually and constantly, but only a few self-builders respected the time schedule and the workload that had been agreed upon:

> The project was difficult. Who can work for two years without taking a day off and spend all the time on-site? So one Saturday, one self-builder did not come. During the week, most of them could not be available. (. . .) And the project fell into chaos.
>
> (Self-builder 11, Ghanaian, Casalmaggiore)

In this context, effective support and guidance were described as fundamental element of the self-building process. Appreciation was expressed for the role of the
mediator in managing conflicts and in calling self-builders to their duty:

> He was good in getting us going. (. . .) He was often among us, he was on-
> site, he monitored how our work evolved. (. . .) Whenever a problem arose,
> he strived to find a solution.
>
> (Self-builder 2, Italian, Casalmaggiore)

> She always did her best and I think she really helped a lot, by trying to pre-
> vent any conflict, by encouraging us to talk and confront each other all the
> time, without leaving any tension distracting us from the works.
>
> (Self-builder 1, Venezuelan, Paderno)

Apparently, by ensuring that self-builders would do their job, interdependency
was facilitated, so that solidarity could develop, in spite of emerging forms of individualism. As such, the introduction of a mediation unit in a self-building project
represents a rather innovative practice in itself, in the wider context of integrated
housing management, and it did contribute to pacific cohabitation and collaboration (see Golinelli, 2006). Yet, the mediators could not always find adequate
strategies to address disagreements and complaints reported by self-builders:

> In spite of mediation, some people did not respect the work schedule. (. . .)
> At the end, those who worked eventually had to accept the situation as it was.
>
> (Mediator, Italian, Casalmaggiore)

Self-builders' opinions were also much more critical about the technical staff and
the representatives of the building society than about the mediator. Dissatisfaction
and disappointment were also expressed with respect to financial management,
particularly for the lack of transparency throughout the project. This made self-
builders feel they were not really part of it:[5]

> He [the project leader] was entrusted with responsibility to define the weekly
> program, he should have been on-site at least once a week but often he was
> not there (. . .) because they had more ongoing projects.
>
> (Self-builder 2, Italian, Casalmaggiore)

> [As it emerged from the narratives of self-builders], they kept on spending
> money (. . .) without knowing if there was sufficient money. (. . .) Then, at a
> certain stage, they stopped paying the providers and the construction material
> did no longer arrive as a result.
>
> (Mediator, Italian, Casalmaggiore)

Following the first year, delays in the delivery of materials started accumulating
on both self-building sites. The situation worsened over time, until the construction process came to a halt. It was at this stage, in March 2010, that Alisei

declared bankruptcy and left the projects. More interviewees, including the Paderno municipal officer, the mediator in Casalmaggiore and the then president of Alisei admitted that the projects suffered from poor financial management. A few self-builders also complained about scarce monitoring by the respective local authorities, which apparently preferred to delegate responsibility for it to the building society. Additionally, following elections and the formation of new governing coalitions, the respective municipalities no longer engaged in the projects. After Alisei went bankrupt, the Cooperative Sottoiltetto started managing the project in Casalmaggiore, while Aler took up the project in Paderno Dugnano.

The emergence of mental and physical fatigue, repeated and growing absenteeism, de-motivation and a sense of abandonment (on the side of institutions) further challenged the solidarity bonds that had been developing among the self-builders, irrespective of their geographical origins:

> On the thirteenth/fourteenth month, tensions started emerging. They were determined by family units. I will give an example. X comes home from a tiring Sunday of construction works, with an aching back. His wife says: When will these houses be finished? I cannot cope with this anymore. I go shopping by myself, I go and pick up the children, I take them to the swimming pool. (. . .) You are the silliest in the group. I have seen Y in the supermarket. Why was he not working? (. . .) There were a few workaholics who fell in love with the project and pushed the group forward and continued working, but then they started feeling resentment towards those who were not devoted to the cause as much as they were.
>
> (Self-builder 1, Venezuelan, Paderno)

Solidaristic relationships began deteriorating and reciprocal trust was affected as a result:

> I have a lot of strong ties, but not with everyone, because with some people we have 'broken up'. (. . .) Had we avoided these things before [absenteeism], relationships would still be strong.
>
> (Self-builder 8, Moroccan, Casalmaggiore)

Initially there was full predisposition by self-builders to meet and share the construction experience as well as to socialize with all the families and their members. Yet, such solidaristic attitudes gradually evolved towards more individualistic ones:

> The bonds that kept us united have faded away a little, in the sense that everyone has their own life now.
>
> (Self-builder 1, Venezuelan, Paderno)

Growing individualism was particularly evident in the scarce engagement of self-builders towards the completion of the common areas and in the conflicts that emerged in the definition of the criteria for the assignment of flats:[6]

We have stopped worrying about getting the common garden finished. Now everyone is mainly concerned about their own house.

(Self-builder 1, Venezuelan, Paderno)

At a certain stage we started discussing whether those who had worked more should have had priority in the choice of their house. The project leader thought this was the right way to proceed (. . .). My family and I thought we had the right to choose and we did so.

(Self-builder 2, Italian, Casalmaggiore)

Individualistic drives became even more evident following the assignment of flats:

Now that everyone has got their own house, they all defend their own space. They do not want their property to be invaded, and this was never like this. (. . .) But I hope the collaboration among us will continue (. . .) I hope it will continue because it would be an advantage for all of us.

(Self-builder 7, Italian, Casalmaggiore)

Now everyone has got their key and they can lock themselves in their own house. (. . .) But I would like things to be like this even in the future. Like you go to someone's house and ask for a hammer, this or that. (. . .) I would like to continue playing, talking, being together with others. (. . .) This is why the municipality gave us the land and we have to continue in this direction.

(Self-builder 1, Ghanaian, Casalmaggiore)

Other self-builders were less positive and expressed fear with respect to growing individualistic drives and the way they could act to dissolve existing bonds:

I have heard various people saying: I will finish my house and then I will close the gate (. . .) In my opinion, at the beginning each one will think for himself/herself. Then let's see over time what happens. . . what happens with the children for example.

(Self-builder 8, Moroccan, Casalmaggiore)

The above suggests that solidarity has been challenged throughout the project. While the actual self-building project has certainly contributed to the emergence of place-based forms of *hic et nunc* (here and now) solidarity, it turns out that such forms of solidarity do not necessarily endure in the long run. On the one hand, it seems hard to hold the community together when the missing 'glue' of interdependency is missing. On the other hand, it is also clear that *hic et nunc* solidarities require external forms of support. In both self-help projects discussed in this chapter, mediators took up this a role. They worked to ensure that all individuals would live up to their commitment and adhere to the rules that have been shared and agreed.

Conclusion

Associated and assisted self-building projects, such as the ones analysed in this chapter, do differ from spontaneous self-building practices that were more common in the past. The latter are based on a form of solidarity that is strongly anchored in pre-existing solidary bonds, shared norms and values and strong feelings of community; the former are characterized by the challenge of stimulating *hic et nunc* forms of solidarity among people who do not share much more than the project they are engaged in. In associated and assisted self-building, people who do not know each other, people with different geographical origins, cultural backgrounds and experiences, suddenly find themselves facing an endeavor in a context of 'solidarity among strangers' (Habermas, 1981). While sharing time and resources, they concede each other the right to remain strangers and to be respected as such.

Through the analysis of two self-building projects in Northern Italy, this contribution has reflected upon solidarity among strangers and how it can operate. Mutual relationships of help and care developed among an ethnically diverse group of self-builders, both in Paderno Dugnano and Casalmaggiore, well beyond the main interest of each individual to build their own dwelling. These relationships were mainly based on an awareness of their interdependency, relating to the heterogeneous and complementary tasks and skills required for the on-site works. Encounters, both on-site and during social events, further fuelled the growth of solidaristic attitudes and practices.

Yet, the empirical analysis has also shown how solidarity was challenged throughout the various phases of the self-building projects. Most importantly, it has demonstrated that there is no guarantee that existing solidarity can consolidate over time and keep the community together. This largely depends on the type and quality of relationships among self-builders and their family members and on the ways conflicts are managed. This points to the fact that *hic et nunc* forms of solidarity, such as those described in this chapter, require some institutional support. Such support should take the shape of an adequate project monitoring, to grant effective implementation, rather than delegation of the tasks to the building society; it should also result in concrete action by the building company, through the help of the mediator, to encourage self-builders to take ownership over the project while adhering to the set rules and honoring their commitment. If properly adopted, these mechanisms can be crucial elements to rethink more effective self-building practices and to sustain a process that evolves out of spontaneous and voluntary encounters (see also D'Amato, 2009). When their adoption is poor, solidarity bonds can weaken instead.

Acknowledgements

This piece of research has been mostly self-funded. Nesta UK, TEPSIE, the Social Innovation Exchange and the University of Oxford Saïd Business School provided a small research grant to support the work on Paderno Dugnano. Earlier versions of this chapter were discussed in various meetings of *COST* Action IS1102 and thanks are due to its members. Special thanks are due also to Roberto Crose, undergraduate architecture student (Politecnico of Milan) who researched on self-building in Paderno Dugnano and to Prof. Tosi (Politecnico of Milan) for his useful insights on housing and to all the people we interviewed in Paderno Dugnano and in Casalmaggiore.

Notes

1 This section draws largely from Semprebon and Vicari Haddock (2016).
2 Objective criteria included: young married or unmarried family units with children; citizenship of an EU Member State or a valid permit to stay; regular residency status or else stable employment in the project's municipality; maximum income between 15,500 and 50,000 euro; and, in the case of Casalmaggiore, no home ownership and no public benefits to buy a house. Subjective criteria were assessed through an interview with self-builders, carried out by a commission of lawyers, sociologists, psychologists and institutional actors to verify predisposition to team and manual work, availability of free time to dedicate to the project, readiness to respect the project's timetable, and possession of manual skills.
3 Only two self-builders were interviewed in Paderno. By the time they were asked for availability, enthusiasm for the project was very low and they were no longer willing to discuss it. Despite repeated insistence, only two self-builders eventually accepted to be interviewed.
4 The research in Paderno Dugnano was carried out by Michela Semprebon in 2012–2014, as part of a postdoctoral piece of research. The research in Casalmaggiore was carried out by Martina Valsesia from 2008 to 2010, as part of her undergraduate thesis (Valsesia, 2010).
5 Empirical evidence on this aspect is limited. Therefore, we will not dwell further on it, although the financial difficulties that emerged at the building sites did influence the motivation of self-builders considerably and this would deserve more reflection.
6 What must be stressed is that this happened, at least in the case of Paderno, in coincidence with the urgency felt by some self-builders to have their flat finished because an eviction was pending upon them. Hence there was certainly more than individualism to similar reactions.

References

Alietti, A., & Agustoni, A. (2013). *Integrazione, casa e immigrazione. Esperienze e prospettive in Europa, Italia e Lombardia [Integration, housing and immigration. Experiences and perspectives in Europe, Italy and Lombardy]*. Milano: Fondazione Ismu, Regione Lombardia, Ministero del Lavoro e delle Politiche sociali.
Allen, C., Camina, M., Casey, R., Coward, S., & Wood, M. (2005). *Mixed tenure, twenty years on: Nothing out of the ordinary*. York: The Joseph Rowntree Foundation.
Amin, A. (2002). Ethnicity and the multicultural city: Living with diversity. *Environment and Planning A, 34*, 959–980.
Arthurson, K. (2012). *Social mix and the city*. Collingwood, VIC: Csiro Publishing.
Atkinson, R., & Kintrea, K. (1998). *Reconnecting excluded communities: The neighbourhood impacts of owner occupation*. Edinburgh: Scottish Homes.
Augé, M. (2002). *In the metro*. Minnesota: University of Minnesota Press.
Baldini, M., & Poggio, T. (2014). The Italian housing system and the global financial crisis. *Journal of Housing and Built Environment, 29*, 317–334.

Ball, M. (2006). *Markets and institutions in real estate and construction*. Oxford: Blackwell.

Banca d'Italia. (2013). Data available on the website: www.bancaditalia.it/statistiche, last accessed 5 September 2013.

Benson, M. (2015). *Talking about self build*. Available at https://selfbuildproject.wordpress.com/author/michaelacbenson/, last accessed 20 August 2015.

Bertoni M., & Cantini, A. (2008). *Autocostruzione associata e assistita in Italia. Progettazione e progetto edilizio di un modello di housing sociale [Associated and assisted self-building. Project management and building project of a model of social housing]*. Rome: Dedalo Libreria.

Camina, M. M., & Wood, M. J. (2009). Parallel lives: Towards a greater understanding of what mixed communities can offer. *Urban Studies, 46*(2), 459–480.

Ceragioli, G., & Maritano, N. (1985). *Note introduttive alla tecnologia dell'architettura' [Introduction to the technology of architecture]*. Torino: CLUT.

CGIL (2013). Costi dell'abitare, emergenza abitativa e numeri del disagio *[Housing costs, housing emergency and the disadvantaged]*. www.cgil.it, last accessed 20 December 2013.

Chaskin, R. J., & Joseph, M. L. (2010). Building 'community' in mixed-income developments: Assumptions, approaches, and early experiences. *Urban Affairs Review*, 45(3), 299–335.

Colombo, M., Martellotta, M., & Solimano N. (2010). *L'autocostruzione: una opportunità per il social housing. Sintesi del rapporto di ricerca [Self-building: an opportunity for social housing. Summary of the research report]*. Firenze: Fondazione Michelucci Onlus.

Cresswell, T. (2004). *Place: A short introduction*. Oxford: Blackwell.

Crow, G. (2002). *Social solidarities. Theories, identities and social change*. Buckingham: Open University Press.

D'Amato, M. (2009). *Vecchie e nuove solidarietà [Old and new solidarity]*. Rome: L'Harmattan Italia.

Dekker, K., & Bolt, G. (2005). Social cohesion in post-war estates in the Netherlands: Differences between socioeconomic and ethnic groups. *Urban Studies, 42*(13), 2447–2470.

Dewey, J. (1954). The public and its problems. In J. A. Boydston (Ed.), *John Dewey: The later works, 1925–1953, Vol. 2*. Carbondale: Southern Illinois University Press.

Fera, G., & Ginatempo N. (1985). *L'autocostruzione spontanea nel Mezzogiorno [Spontaneous self-building in the Italian Mezzogiorno]*. Milan: Franco Angeli.

Ferri, G. (2011). *Il Gestore Sociale. Amministrare gli immobili e gestire la comunità nei progetti di housing sociale [The "Gestore sociale". Managing building blocks and managing the community in social housing projects]*. Milan: Altreconomia Edizioni.

Fundarò, A. M. (Ed.). (1977). La dimensione dell'azione: ambiente e costruzione. Partecipazione, autogestione, autocostruzione *[The dimension of action: Environment and construction. Participation, self-management, self-building]*. Palermo: STASS, Faculty of Architecture, University of Palermo.

Galeazzi, C., & Valsesia, M. (2013). *Laboratorio sociale – Verso la definizione di nuovi modelli di gestione residenziale [Social laboratory – Towards the definition of new models of residential housing management]*. Milan: Editore Regione Lombardia.

Golinelli, M. (2006). Buone prassi di housing sociale: il ruolo protagonista della cooperazione sociale [Good practices in social housing: The crucial role of social cooperation]. *Impresa Sociale*, 2/2006, 156.

Goodchild, B., & Cole, I. (2001). Social balance and mixed neighbourhoods in Britain since 1979: A review of discourse and practice in social housing. *Environment & Planning A, 19*, 103–122.

Graham, E., Manley, D., Hiscock, R., Boyle, P., & Doherty, J. (2009). Mixing housing tenure: Is it good for social well-being? *Urban Studies, 46*(1), 139–165.

Habermas, J. (1981). Toward a European political community. *Society, 39*(5), 58–61.

Harvey, D. (2009). *Cosmopolitanism and the geographies of freedom.* New York: Columbia University Press.

Harris, R. (1999). Aided self-help housing, a case of amnesia: Editor's introduction. *Housing Studies, 14*(3), 277–280.

Illich, I. (1974). *La convivialità [Conviviality].* Milan: Arnoldo Mondadori Editore.

Jessop, B., Brenner, N., & Jones, M. (2008). Theorizing socio-spatial relations. *Environment and Planning D: Society and Space, 26*(3), 389–401.

Joseph, M. (2006). Is mixed-income development an antidote to urban poverty? *Housing Policy Debate, 17,* 209–234.

Kearns, A., & Forrest, R. (2000). Social cohesion and multilevel urban governance. *Urban Studies, 37*(5–6), 995–1017.

Kleinhans R., Priemus H., & Engbersen G. (2007). Understanding social capital in recently restructured urban neighbourhoods: Two case studies in Rotterdam. *Urban Studies,* 44(5–6), 1069–1091.

Kleit, R. (2008). Neighbourhood segregation, personal networks, and access to social resources. In J. Carr & N. Kutty (Eds.), *Segregation. The rising costs for America* (pp. 237–260). New York: Routledge.

Lefevre, H. (1094). *La Production de l'espace.* Paris: Anthropos.

Lepofsky, J., & Fraser, J. C. (2003). Building community citizens: Claiming the right to placemaking in the city. *Urban Studies, 40*(1), 127–142.

Loopmans, M. (2000). Het bedrog van de buurt: Mogelijkheden en beperkingen van residentiële sociale mix in de strijd tegen sociale uitsluiting [The deceit of the neighbourhood: Possibilities and restrictions of social mix in the struggle against social exclusion]. *Agora, 16,* 26–28.

Matejskova, T., & Leitner, H. (2011). Urban encounters with difference: The contact hypothesis and immigrant integration projects in eastern Berlin. *Social & Cultural Geography, 12,* 717–741.

Meroni, A., & Sangiorgi, D. (2011). *Design for services.* Aldershot: Gower Publishing.

Mugnano, S., & Palvarini, P. (2013). Sharing space without hanging together: A case study of social mix policy in Milan. *Cities, 35,* 417–422.

Oosterlynck S., Loopmans, M., Schuermans, N., Vandenabeele, J., & Zemni, S. (2016). Putting flesh to the bone: Looking for solidarity in diversity, here and now. *Ethnic and Racial Studies, 39*(5), 764–782.

Paolella, A. (2008). *Attraverso la tecnica. Deindustrializzazione, cultura locale e architettura ecologica [Through the technique. De-industrialization, local culture and ecological architecture].* Milan: Elèutera.

Scanlon, K., Whitehead, C., & Arrigoitia, M. F. (Eds.). (2014). *Social housing in Europe.* Chichester: Wiley-Blackwell.

Semprebon, M., & Vicari Haddock, S. (2016). Innovative housing practices involving immigrants: The case of self-building in Italy. *Journal of Housing and the Built Environment, 31*(3), 439–455.

Silverman, E., Lupton, R., & Fenton, A. (2006). *A good place for children? Attracting and retaining families in inner urban mixed income communities.* York: Joseph Rowntree Foundation.

Simmel, G. (1950). *The sociology of George Simmel.* Glencoe, IL: The Free Press.

Valentine, G. (2008). Living with difference. Reflections on geographies of encounter. *Progress in Human Geography, 32,* 323–337.

Valsesia, M. (2010). *Autocostruzione associate e assistita: una possibile risposta alla domanda di casa e alle domande sulla casa. L'esperienza di Casalmaggiore in Lombardia [Associated and assisted self-building: A possible solution to the demand on and for housing]*. Unpublished undergraduate thesis, University of Milan-Bicocca.

Van Kempen, R., & Bolt, G. (2009). Social cohesion, social mix, and urban policies in the Netherlands. *Journal of Housing and the Built Environment, 24*(4), 457–475.

Vestbro, D. (2000). *The role of design and planning professionals for solving the global housing problem*. Paper presented at the Conference Challenges for Science and Engineering in the 21st Century, organized by the International Network for Engineers and Scientists (INES), Stockholm, 14–18 June.

6 Challenging the figure of the 'migrant entrepreneur'

Place-based solidarities in the Romanian arrival infrastructure in Brussels

Bruno Meeus

Introduction

In 2004, then Romanian prime minister Adrian Nastase argued that the two million Romanians working abroad contributed to the construction of a new Romania. In the West, Nastase suggested, a new and for the country much-needed, responsible citizen with a new work ethic was being created. In his own words, they 'learned how long a coffee break should take' (Badescu, 2004).

Nastase's speech, now more than a decade ago, is characteristic of a discursive construction of migrants as entrepreneurs that has settled itself in developmental discourses more broadly meanwhile. Viewed as risk-taking subjects, states and development institutions such as the World Bank increasingly imagine returning migrants as ideal 'development' partners who embed and extend competitive market rationalities into everyday social relations and institutions (Mullings, 2012, p. 407). This image of returned citizens who have re-educated themselves through experiences abroad has important depoliticizing consequences. In Romania, it leads attention away from the responsibilities of the Romanian government to provide welfare for all citizens (Meeus, 2013). In the city of arrival, it does not address or challenge the atrocities of exploitation migrants often experience, but treats them as useful learning moments.[1]

This broad *problématique* of the rise of a meritocratic morality that constructs migrants as entrepreneurs and considers social mobility as an individual do-it-yourself responsibility is the starting point of this chapter. Drawing on fieldwork in Brussels, I am interested in the innovative forms of solidarity that challenge such a morality. For a decade, the multi-scalar legislative context regulating access to the labour market in Brussels has left precarious self-employment as one of the few channels for Romanians into the legal job market. Against this background, I explore two cases in which 'sleeping' discourses, practices and infrastructures of solidarity have been reactivated, renegotiated and renewed in particular places in Brussels as a result of the arrival of Romanian newcomers.

The first case describes place-based solidarity in a Romanian migrant church in Brussels and can be referred to as 'ethnic niche solidarity'. Here, the potential mechanisms of solidarity came along with the migrants from Romania and crystallized in innovative but small-scale forms of ethnic- and faith-bound solidarity knotted together by the newly erected church in Brussels. The second case is more in line with the 'solidarity in diversity' argument of this book and consists of a local network of unionized railway employees. Here, the potential mechanisms of solidarity were already in place in Brussels for decades, but were rethought when Romanian self-employed workers shared the same workplace with Belgian colleagues. In both cases, previously existing infrastructures of solidarity were reactivated. While the actions of union members tended to directly challenge the further institutionalization of meritocratic moralities, the solidarity initiatives at the Romanian-Orthodox church tended to reinforce them.

The fieldwork behind this chapter was conducted between November 2013 and December 2015 in the framework of a more general three-year Innoviris postdoctoral research in which I compared the infrastructures of arrival for Romanian and Bulgarian migrants in Brussels. Arrival infrastructures are those parts of the urban fabric where newcomers become entangled upon arrival and where their future local or translocal social mobility is produced as much as negotiated (Meeus et al., forthcoming).

To sketch the general picture of incorporation of Romanian migrants in Brussels, I draw on secondary literature, 43 short interviews (around 10 to 20 minutes) and five in-depth interviews (more than two hours) with Romanian migrant workers. The short interviews were mainly conducted near the weekly open air market in the neighbourhood of Kuregem[2] (30), in front of the Orthodox church (10) and in the central train station (three). The longer interviews were conducted in the homes of the respondents with the help of an interpreter. My comparison of the two place-based solidarity initiatives is also based on different interviews with religious staff (three in-depth interviews with a priest), regular meetings with railway employees and articles in newspapers and magazines.

The chapter proceeds along four more sections and a conclusion. The following section summarizes the main arguments in the 'solidary ethnic niche' literature. It highlights that the performance of ethnicity abroad does not necessarily only lead to forms of solidarity, but can foster meritocratic moralities as well. The third section gives a short overview of the multi-scalar policies that funnel Romanians in Brussels into precarious self-employment, the risks these self-employed Romanians run and the two grammars that institutional actors deploy to tackle these risks. The chapter then moves on to an analysis of the two case studies and ends with a conclusion.

Ethnicity and situational we-ness: between solidarity and meritocracy

In migration research, the idea of the 'solidary ethnic niche' as a form of incorporation for migrant newcomers takes an important place. A range of studies has shown that the establishment of ethnic niches in the labour market occurs as a result of a combination of 1) institutional and legislative preconditions that confine migrant groups to particular labour market niches; 2) discrimination and other forms of informal social closure that block access to the labour market; 3) gaps in the labour market others have left vacant; 4) social networks through which information circulates and the recruitment of co-ethnics takes place; and 5) migrants' preferences to work along co-ethnics and in particular labour market niches (see among others, Castles & Kosack, 1973; Portes, 1995; Waldinger, 1995; Kloosterman et al., 1999; Rath, 2002; Schrover et al., 2007).

Building upon this work, Portes and Sensenbrenner (1993) have developed a more expanded view on ethnic niches that includes intra-ethnic solidarity beyond access to employment. A crucial point in their argument is that 'ethnically bounded solidarity' abroad is a form of situational solidarity that mobilizes pre-existing sources of solidarity: it draws on either previously existing social capital that resides in clan, family and kinship relations or on a situational (re) construction of national, ethnic, religious or hometown group identities. Drawing on Marx's description of class consciousness, Portes and Sensenbrenner (1993, p. 1328) argue that bounded solidarity abroad develops among 'members of a particular group who find themselves affected by common events in a particular time and place'. This sentiment of 'situational we-ness' can activate 'dormant feelings of nationality among immigrants' and create 'such feelings where none existed before' (Portes & Sensenbrenner, 1993, p. 1328), as was the case for Sicilian peasants in New York who learned to think of themselves as Italians (Glazer, 1954, cited in Portes & Sensenbrenner, 1993). Hence, their account highlights that previously existing social networks only emerge as an infrastructure of solidarity because of the immigrant context and that the imagined national, ethnic, religious or hometown community is only feeding solidarity because of a sentiment of situational we-ness that draws on these imaginations.

Recently, researchers have started to debunk the myth that ethnicity is always a resource for solidarity abroad. In fact, over the past decade, a number of scholars have argued that migration research should not take nationality or ethnicity for granted (Glick Schiller, 2008; Fox & Jones, 2013), but approach it, instead, as a contextual performance (Butler, 1988; Sullivan, 2012). The question, then, becomes why and how actors 'act ethnic' through speech and discourse, through their bodies and material practices, consciously or not (see Levin, 2014). A careful look at performances of ethnicity will help us, I would add, to analyse where and when ethnicity enables situational solidarity to grow and where and when it entangles with individualist meritocratic and competitive values and practices to produce forms of situational meritocracy.

In his research on Filipino seafarers, McKay (2007) demonstrates, for instance, how a particular ethnic niche is maintained through masculine and national/ethnic performances at home and on board. These performances help to endure the subordinate position of the Filipinos on board, while neither the exploitative relations nor the blatant racism are challenged. The imagined we-ness (us, Filipino seafaring men), therefore, feeds a competitive disposition, rather than a solidary one:

> Filipino seafarers have [. . .] constructed transnational meanings around their work and masculinity that help them invert – or at least obscure – their secondary or marginal status [on board]. Their conspicuous consumption, their

reputation as hyper-masculine adventurers, and their ability to endure hardships, all give Filipino seafarers a chance to transform a marginalised and subordinate masculinity on the job into a model of exemplary masculinity at home, emphasising the ideals of fatherhood, economic provision, sacrifice for one's family, and the 'machismo of manual work'.

(McKay, 2007, p. 630)

Recent research in the UK has documented similar forms of ethnic performances and situational meritocracy in encounters between New Member States (NMS) migrants and a diversity of other workers with whom they compete in particular segments of the labour market. On construction sites, for instance, Polish workers tend to distinguish themselves from British ones by depicting the latter as lazy and welfare-dependent and by emphasizing their own Polish ethic of hard work instead (Datta, 2009; Datta & Brickell, 2009; Fox et al., 2015). In the context of domestic work, Kilkey and Perrons (2010) document how NMS migrant handymen build on the existing stereotype of the 'Polish Plumber' to create an ethnic-masculine brand of 'Polish Handyman'. By presenting themselves as members of the host country's 'white' majority, Polish (and Lithuanian) migrants in London put themselves in a powerful position in relation to other migrant groups in the city (Parutis, 2011). Similar constructions of whiteness play an important role for Romanian workers in the UK. Moroşanu and Fox (2013) demonstrate, for instance, that many Romanians do not fight discrimination on the labour market through a positive reappraisal of 'Romanianness', but through a transfer of the negative stigma onto the ethnic Roma with whom they are frequently associated.

In the discussion that follows, I will contrast two complex forms of situational solidarity that grew in Brussels. In the first example, the Romanian church reactivates and knots together a range of small-scale infrastructures of solidarity that relieve some of the atrocities of Romanian workers' life in Brussels. It also reveals, though, that a situational sense of 'shared misrecognition' reproduced by members of the church feeds an ethnicized meritocracy. The second example demonstrates how a long-established large-scale infrastructure of solidarity – the trade union – was reactivated in a particular workplace. Regular encounters between unionized white Belgian employees and non-unionized Romanian self-employed resulted in the construction of a situational we-ness beyond ethnic-meritocratic performances, so that the solidarity infrastructure of the trade union eventually included the Romanians as well.

The Romanian 'entrepreneur' in Brussels

Rath and Swagerman (2011) argue that only few European countries and cities have developed specific policies encouraging 'ethnic entrepreneurship'. While the right to move freely, to stay and to work in another member state – the so-called 'freedom of movement of workers' – is one of the four pillars of the European Common Market, the 'freedom to establish a business' remained one of the few channels into formal work abroad for a decade for many NMS citizens. As 'Old

Europe's' national electoral agendas dominated the conditions of intra-European mobility for citizens of the NMS that joined the EU in 2004 and 2007, a European-level agreement allowed national governments to install a 2+2+3 transitory period (Tamas & Münz, 2006; Krings, 2009).

In 2004, 12 of the 15 pre-enlargement EU member states, including Belgium, used this agreement to restrict access to their labour markets for citizens of new EU member states. Some countries like Austria, Belgium, Italy and the Netherlands offered labour permits on a quota base. The United Kingdom, Ireland and Sweden did not implement such restrictions, but access to certain social welfare bene-fits was curtailed in the UK and in Ireland (Moriarty et al., 2015). In the second period, the option of a transitory period was installed again at the European level to counteract increasing uneasiness in public opinion (Pijpers, 2006), an oppor-tunity now taken by UK and Ireland to restrict Romanian and Bulgarian migrant workers access to their labour markets as well. These restrictions were removed on 1 January 2014.

The case of NMS migrants in Brussels reveals that entrepreneurship in the form of self-employment is the result of European Union and national policies that leave self-employment as the last resort for migrants. In the Brussels Capital Region, the number of NMS migrants has risen sharply over the last decade. Out of a total population of about one million people, the number of Romanians, Poles and Bulgarians has increased from 1,978, 5,489 and 825 in 2004 to 29,682, 26,414 and 9,746 in 2014, respectively (Hermia, 2015). Figure 6.1 suggests that most of these NMS migrants are self-employed.[3] Between 2007 – the moment of EU accession for Romania and Bulgaria – and 2014 – the moment when restrictions were lifted – the two most important options to escape the circuit of unregistered work for Romanians in Belgium were 1) starting up a business and 2) working in one of the 'shortage jobs' for which the employer had to demand a permit B.[4] Yet, when restrictions for Polish workers were lifted in 2009, the number of self-employed Polish stabilized.

Even though self-employment offers a route towards social security in Belgium, the status is clearly not without risks. First, while access to work in the UK and the Netherlands occurs mostly through formal channels like temporary staffing agencies, most of my respondents in Brussels gained information about job opportunities through kinship and friendship networks. In many cases, this also led to unregistered work. Second, different respondents working in construc-tion reported about the difficulty to bridge the periods without income between jobs. A much-used strategy was to pool household income. When the male part-ner registered as self-employed, his female partner could more easily obtain a permit and find employment, for example through the service voucher system (*dienstencheques*) in the cleaning sector. Third, different self-employed respond-ents reported a history of accumulating debts. Typical examples are construction workers who work officially as 'business associates' rather than as employees, but do not receive money from the 'patron'. Other instances include sellers in grocery stores or cleaners who are registered at one of the social security offices for the self-employed, but who do not know they have to pay social security

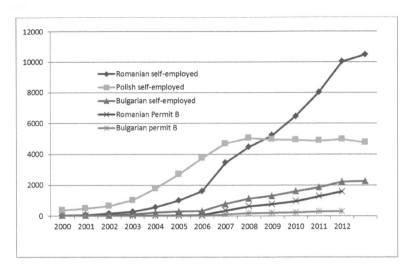

Figure 6.1 Romanians, Polish and Bulgarians on the Brussels Capital Region labour
market: self-employed and permit B (2000–2013).

Source: RSVZ; due to limited availability of data, the number of Romanian and Bulgarian permit B
holders is based on extrapolation of the 2012 data.

contributions on a regular basis. When the social security office confronts them
with the accumulated arrears, they face a debt mountain difficult to repay. While
the RSVZ (Social Security for Independent Entrepreneurs) have no data about the
number of self-employed Romanians currently indebted to the social security, it is
known that the total debt amounted to about one billion euro in 2012.

In institutional settings, two different solutions to the vulnerable situation of
Romanian migrant workers in Belgium are proposed. As the self-employment
statute and the posted-worker system neatly fit the tendency to phase out the regular employee contracts in a number of employment sectors – a tendency against
which labour unions protest – it is no surprise that Belgian unions, together with
organizations like OR.CA (Organization for Clandestine Labour Migrants in
Belgium) and the Centre for Equality of Chances and the Fight against Racism,
advocate a structural approach that deals with the legislative conditions which
funnel newcomers in risky employment situations. Towards the end of the second transitory period, the cabinet of the Minister of Work, Monica De Coninck,
argued for a new extension because 'the Belgian labour market faces an ever
greater pressure as a result of the financial and economic crisis' and because 'we
have to avoid to add more factors that could potentially lead to a situation even
less manageable' (cited in Danckaers, 2012). The coalition of unions and other
organizations argued, however, that the number of migrants was not the problem,
but the lack of strong legislative tools. Without success, they made a strong plea
for ending the labour market restrictions for Romanians and Bulgarians and for
focusing on the fight against social fraud instead.

In recent years, a second approach has been gaining traction. It emphasizes the potential of regulated entrepreneurship and the legal status of self-employment as channels out of unregulated work. The integration policy of the Flemish Community Commission – the policy level which represents the stakes of the Flemish community in Brussels, which has its own migrant integration policy next to that of the French community – has started to stimulate and regulate self-employment of newcomers, for instance (Bourgeois, 2009). In 2012, the Romanian self-organization Arthis (Belgian-Romanian cultural house) became a partner of Flemish integration policy with a project called '*van immigrant tot middenstand*' (from immigrant to small entrepreneur). The aim of the project was to lead Romanians working irregularly towards the formal status of self-employment and to inform them about the procedures and the advantages of registering as self-employed. In the framework of the project, different information sessions were organized, including an evening in which the possibilities of micro-credits were presented.

Now that I have explained that the growth of Romanian 'entrepreneurs' in Brussels is the result of multi-scalar legislative frameworks which leave them few other options, I will move to my two case studies in Brussels to explore to what extent place-based solidarities potentially challenge the meritocratic morality of 'do-it-yourself' entrepreneurialism, a morality that depoliticizes these legislative frameworks.

Solidarity and ethnicized meritocracy in a migrant church

Together with the rising number of Romanians, Romanian-speaking church services for Orthodox, Greek Catholic and plenty of (neo-) Protestant denominations (Adventists, Baptists, Pentecostals) have been established in Brussels. As demonstrated by among others Trappers (2009), church services are one of the few more elaborate forms of Romanian self-organization abroad.

In 2005, when existing Romanian Orthodox church services were bursting at the seams, a desacralized Catholic church was bought by the Romanian Orthodox patriarchate. With the financial help of the Romanian state and Romanian company managers in Belgium, the church was restored by volunteers. On a regular Sunday, about 500 people attend mass. On special occasions, like the Easter celebration, this number rises to up to 4,000 churchgoers.

According to Ley (2008), migrant churches offer newcomers a place to meet fellow immigrants with shared existential concerns (living in a foreign city, the trauma of a foreign language, difficulties finding a job) and shared biographies (migration from the same region, country). Ley's account suggests that these shared concerns and biographies form the basis for solidarity to develop among the faith community and that the social capital accumulated in migrant churches enables them to be urban service hubs as well as spiritual homes.

Recurring interviews with the priest of the newly established Romanian church confirm that the church currently forms the node of several solidarity mechanisms. A first mechanism of direct solidarity draws on pre-existing social networks in villages and regions of origin. Asked about the grounds on which a man with thousands of euros debt was helped, the priest emphasized the importance of two particular village communities in Brussels:

> These two communities are so important, so compact. The people here, from Maramures, they all mobilized not to leave him to his fate.
>
> (Priest)

In case of huge debts (the priest has different examples of debts over 50,000 euro) it appears that the reputation of the person in need is of quintessential importance:

> *Bruno:* Is it also your task to signal problems and to assure the faith community that a particular person in need is trustworthy?
>
> *Priest:* Well, in this situation, and most of the time, they already have a certain reputation, they are employers, they are not people anonymously working for an hourly wage and nobody knowing them. They are people who work with dozens of companies. In this case, there were 50 employees, which means 50 families behind him, who all have families and friends. I mean, I'm not the one constructing his reputation. He is already known. My work is rather to support these people emotionally.

While these networks already existed before migration to Brussels, I want to emphasize the importance of the church as a central node. Regular religious practices (attending Sunday's mass, baptisms, Easter) help to maintain social networks and to maintain reputations. These practices also enable the creation of new social capital, as people who do not come from the same region of origin or a similar social class come to the church as well. Just like a Korean church leader in Ley's (2008, p. 2063) study stated that 'the church is the "knot" of different strings coming together, the centre', the priest claims:

> All social profiles are here. We are a 'little Romania', all categories, housewives, self-employed, university researchers, people working for the EU, everyone. But I think, around 70 percent work as self-employed, in construction work and cleaning.
>
> (Priest)

A second mechanism of more indirect solidarity in the church has been set up by the priest. In only a short period of time, he has managed to develop a small-scale solidarity infrastructure that draws on mainly two resources: 1) the religious ritual of offering a financial contribution for the seven works of mercy or 'Misericordia'; and 2) the voluntary work of members of the faith community.

The Misericordia fund is managed by an association chaired by the priest. Money is not being lent to people, but indebted members of the faith community can receive a voucher from the association that manages the fund and exchange it for a free food package or for cleaning supplies from Romanian-speaking vendors in supermarkets and grocery stores in Brussels. The vendors can reclaim the money from the association with the voucher. According to the priest, this system can only work with Romanian vendors because 'a certain level of trust is needed'. After all, the voucher-system was set up as a result of tensions around who deserves the solidarity of the church community:

> Without prejudices, but people in need could spend this money on liquor. (. . .) From the start we tried to help, but there were people taking advantage of it, who kept coming every day. And then it causes a heartache to say 'I don't give you anything', because the person is suffering. But at the same time, I cannot be irresponsible. So I had to find a solution in order for the help to be correct and ethical, especially towards the people who give money within our system of solidarity collection. That's also why we created an association.
>
> (Priest)

Apart from the fund, the motivation to 'do something for the community' has also fuelled various forms of volunteering such as language and literacy courses. To deal with Belgian bureaucracy and legislation, the priest has even established a 'knowledge community' (Blommaert, 2011) or a local centre of expertise specialized in what the priest calls '*se débrouiller en context*' or 'making ends meet in context'. Romanian transport companies also offer bus tickets at reduced prices for members of the faith community who lost all their financial means and want to go back to Romania.

All in all, the 'little Romania' around the Romanian church in Brussels can be considered as a form of 'situational we-ness'. By mobilizing the idea of an ethnic-religious (Romanian-Orthodox) commonality in the Brussels context, it is discursively created by people such as the priest. As the faith community includes a diversity of occupations, this commonality transcends social classes that usually divide Romanians abroad (see Trappers, 2009). Nonetheless, the imagination of an Orthodox-Romanian nationhood in Brussels influences the struggle for the symbolic incorporation of the Romanian self-employed in Brussels.

While Polish workers in Brussels have established a strong reputation of trustworthy cleaning ladies and heavy drinking, but hardworking handymen (Paspalanova, 2006), Romanians struggle with their symbolic incorporation in the city. In her comparative research on the reputation of Romanians in Lisbon, Brussels and Stockholm, Trappers (2009) demonstrates, for instance, that Romanian migrant workers in Brussels are dominantly seen as poor, gypsies and criminals, and only to a limited extent as trustworthy workers. Pointing to the social class differences between different Romanian migrants, this image is sometimes even shared by Romanian expats (Trappers, 2009).

Similar to what Moroşanu and Fox (2013) report, Romanians in my research repeatedly reattached the stigma to an ethnicized other. A recurring phrase I registered during the interviews I conducted at the market was that 'the Roma give use a bad image in Belgium'. Other ethnic constructions served the same purpose:

There are lots of Arabs here, they don't work, they are on social benefit.

(Marco,[5] 40, self-employed construction worker)

Belgian workers are fine, but they are not hard workers. When they've had a few pints, they don't show up.

(Mitu, 42, self-employed meat worker)

It is important to know this context to understand why the Romanian priest stages the social needs of the Romanians in Brussels in a particular way. As Wills (2008) has pointed out, faith-based organizations that reach undocumented and exploited workers can form solidary coalitions with labour unions to campaign against exploitative labour conditions in a particular place. There is certainly potential in the Romanian church to develop such campaigns. 'Organizing one's social security' is one of the important services offered by the support network. The housing problems of many Romanian workers are also recognized as a problem other groups in Brussels experience as well (Meeus & Schillebeeckx, 2015). The priest's discourse reveals an important pitfall, however. To the outside world, the priest is tempted to sketch a positive image of the 'Romanian community' by emphasizing the ethnic differences between Roma and Romanians and by purifying non-Roma Romanian we-ness in Brussels:

Unfortunately, the image of the Romanians in Brussels is not good. The stupidities of some receive much more attention in the media. Begging gypsies are identified with the Romanian community, while they belong to a different ethnicity.

(Priest)

Central to this discourse is the success of the Romanian entrepreneur in Brussels:

I now understand why the Unites States are leading the way. There is a discipline among migrants. Our Romanians do not leave home to go on vacation or to work only eight hours a day. No, they work ten or twelve hours a day. They accept activities others do not accept, take contracts others do not accept. The success of Romanian enterprises – and there certainly is a success, particularly in construction – is that they solve problems others cannot solve, they respect the planning, they work on Saturdays, ten hours a day, not eight. (. . .) According to my information, there are a lot of Romanian enterprises with a very good reputation, in construction but in cleaning as well.

(Priest)

As such, my analysis hints at a combination of bounded small-scale solidarity infrastructures and ethnic moralities of meritocracy that both draw on place-based sentiments of we-ness. In order to determine who is eligible for solidarity, a boundary is drawn between 'pure' and hardworking Romanians, on the one hand, and ethnic and non-entrepreneurial others, on the other. While this border is probably not closely guarded in practice, it lives on in the speeches of the priest and the stories of failure and success shared between members of the faith community. In this way, the pedagogical environment of the church teaches the churchgoers that successful migration signifies becoming a hardworking entrepreneur. Romanian entrepreneurs who have made it offer a role model for newcomers, a social position to which to aspire.

Unionized solidarity in diversity

The degree of unionization at the Belgian national railway company (NMBS) is traditionally high. According to Jos Digneffe (2008), then chairman of the Flemish wing of the socialist railway union (ACOD-Spoor), around 86 per cent of railway workers are unionized. In comparison, the average degree of unionization in Belgium is estimated to be around 60.5 per cent (Vandaele & Faniel, 2012).

Over the past two decades, the railway union has tried to defend the institutionalized railway employee status during successive liberalization campaigns. When European directives to liberalize large public infrastructures urged the Belgian railway company to split up in two entities in 2005 (Infrabel manages the infrastructure, NMBS organizes the train service), the unions managed to defend the railway employee status through the creation of a third company, the NMBS-holding, which organizes the human resources for both companies and the management of the major train stations.

Yet, to circumvent the costly railway employee-status, the NMBS-holding started outsourcing activities that were not considered to be core tasks of the public transport company. Private companies can submit an offer to clean a particular workplace for a particular period, for example. Extra services, such as language courses, can make their bid more attractive, but the directors generally prefer the cheapest contract (Guldentops, 2014). As a result, Belgian railway employees working as cleaners, directly employed by the NMBS-holding and unionized in the railway union, now work alongside colleagues, most of them not unionized and non-Belgian, who have temporary contracts and receive lower wages to do the same work. In train stations, different private cleaning companies employ different workers, often through subcontractors or temporary staffing agencies.

In what follows, I will investigate whether members of the railway union – not necessarily cleaners – can connect workers with different status in different workplaces. By way of example, I draw on the experiences of seven Romanian 'entrepreneurs' who had started working as self-employed cleaners during the nightshifts in some of the Brussels train stations in 2007. After tracking down three of these workers in 2015, I asked them why they had chosen the cleaning sector in the first place. They replied that they had not really been offered a choice. In the construction sector, where they had been working before, the uncertainty was simply too high. The Belgian Turks who owned the construction firm that employed them only paid their salaries after three months. When they started looking for better jobs, Romanians from another part of Romania, who worked on the same construction site, told them about vacant places in the night shift of the cleaning team in one of the Brussels train stations, which they took up.

[In construction], you work two weeks and you don't work the next two weeks and you don't know when there will be work. In cleaning, there is at least a certainty of income, although we had to cope with eight day working weeks without a day vacation, without health insurance.

(Ioan, self-employed cleaner)

In 2008, one of the Romanians was approached by a unionized Belgian cleaner working for the NMBS. As a railway employee, he had been working side by side with some of the Romanian 'entrepreneurs' at night. As such, he had remarked the sheer impossible number of night shifts his colleagues were running and discovered they worked no fewer than 350 days a year, weekends included. After gaining the trust of the Romanian workers, the unionized cleaner contacted the general secretary of the socialist cleaning union (FGTB) who informed the social inspection. After visiting the workplace, the social inspection concluded that the Romanians worked as self-employed for the transnational cleaning company GOM, but could not even choose their own work schedule. Even though the social inspection had already written a report about the social fraud in 2008, the Romanians were still working under the same exploitative conditions four years later.

After numerous demands to GOM to make the contracts of the Romanians public, the general secretary of the socialist cleaning union decided to put the chief employer – the NMBS – under pressure by contacting the press. Hence, in February 2012, the Flemish job magazine *Vacature* reported about the situation of the Romanians (Soenens, 2012). The report was widely taken up in Flemish- as well as French-speaking press. Eventually, the discussion even reached the national parliament.

As a result of the pressure, the NMBS-holding agreed to recruit 20 new cleaners for the Brussels train stations with a prospect to receive a regular (railway employee) contract. The new employees were long-term unemployed people that took part in an activation programme. Once the contract of GOM had expired in 2013, some of the Romanians were employed by a new Turkish cleaning company, but only during weekend night shifts and in different train stations. During the week, they had started working as unregistered construction workers again.

In a reaction to their vulnerable situation, a group of railway union members consisting of cleaners, technicians and others sharing the workplace with the Romanians started an information campaign on trains passing through Brussels, together with a petition to demand a regular contract for the Romanians. The petition, written by one of the railway employees, framed the struggle as a shared struggle against new attacks on the railway employee status. After describing the story of the Romanians, the text states:

In this case, subcontracting is a way to by-pass the statute of the railway employee and the regulations of the NMBS, and to literally break down the public services. NMBS-employees often have difficulties to be in solidarity with labour migrants who take away 'their jobs', but this solidarity is, however, the only possibility to ameliorate the labour conditions of all the railway employees, whoever their legal employer may be.

(Guldentops, 2012)

While one of my Romanian respondents reported minor incidents of bullying and ignorance by Belgian colleagues in the shared kitchen and cloakroom in one of the Brussels train stations, the rapprochement in another train station in Brussels during the night shifts created a first creative moment of union action.[6] Through the horizontal networks among railway employees in different train stations, this rapprochement eventually spilled over to other workplaces. The story of the Romanians became an exemplary story of victimhood caused by privatization policies, a direct threat to their own jobs. Without following the usual vertical procedures of railway union actions to be legitimated by the top of the organization, the railway employees translated the needs of the Romanians into an already existing repertoire of railway union arguments, so that other members of the railway union could be convinced to sign the petition: the railway employee status was under attack! The Romanians were not the enemy, but the recruitment policies of the NMBS-holding were! In this process, a new situational we-ness developed beyond ethnic lines of division.

The workplace did not only turn out to be a crucial pedagogical environment for the unionized railway employees, however, but also for the Romanians themselves. As 'migrant entrepreneurs' working under the price, they were considered to be a threat to the unions, but as victims of exploitation and 'employees of the NMBS in practice', they became part of a shared struggle. As was the case for the union members, this struggle was not to their own advantage. After all, they ended up losing the job they had preferred over the uncertain work on construction sites. While this uncertain work in construction acted, somehow paradoxically, as a safety net during the course of the actions, at least three of the seven Romanians currently engage themselves in union actions regularly. As such, the struggle for worker rights in the Brussels train stations also created a new Romanian subject, one that did not revolve around individual entrepreneurship and competition, but one centred around solidarity and cooperation.

Conclusion

I have situated this chapter in the context of the rise of discourses on the figure of the 'successful entrepreneurial migrant'. This figure can be found in development policy discourses, where the returning migrant entrepreneur is constructed as a welcome development partner, and to a certain extent also in newly developing urban governance discourses, in which the poor urban immigrant newcomer is reframed as a source of entrepreneurial potential. As this discourse tends to construct social mobility as an individual responsibility, it legitimizes the roll-off of societal risks on individuals, both in places of departure and arrival, and ignores other discourses and practices of collective emancipation.

The case of Romanians who work abroad is a particularly insightful case to understand the causes and the effects of this discourse. While my previous research shows that migration strategies are the result of the individualization of social risks through the rolling out of the Romanian state (Meeus, 2013), my analysis in the third section demonstrates that this model of do-it-yourself social mobility is further endorsed in the context of arrival in Brussels. Between 2007 and 2014, Romanians were funnelled into self-employment as a result of different governmental scales, ranging from the European (the right to establish a business) to the national (restricting free access to the labour market) and the regional (encouraging ethnic entrepreneurship). The fourth section demonstrates that Romanians also reproduce this figure of the successful entrepreneur themselves through discourses of whiteness and 'hardworking'.

In this chapter, I have started to explore how solidarity can challenge – and form an alternative for – the further depoliticization of do-it-yourself welfare. Through two different case studies, I have studied two very different kinds of solidarity. In the second section, I have paid attention to the literature on solidary ethnic niches, in which it is suggested that the experience to find oneself 'thrown together' with compatriots in a foreign context can create a sentiment of situational we-ness that feeds intra-ethnic cross-class solidarity abroad. Confirming research by Trappers (2009), my interviews with Romanians in Brussels demonstrate that the central node of a church, in which construction workers, cleaning ladies and businessmen share the same spiritual place, is central in this process. Shared religious practices enable the emergence of a small-scale infrastructure of indirect solidarity: the voucher system. The beneficiaries of this solidarity are defined along purified ethnic-religious lines, however. Even more importantly, as the priest heralds the entrepreneurial success of the members of his purified ethnic faith community, it appears that this inward-looking solidarity is but a form of compassionate harm reduction. In general, individual success through working harder and more hours and going under the price are legitimized as the order of the day.

Ley (2008) suggests that bounded solidarity in newly established migrant churches is a temporary phase. As new generations grow up with less interest in religion, fewer emotional ties with the homeland and diversifying biographies, the ties that bind the different networks together weaken. According to Ley, over time these migrant churches either shrink and even disappear or make a transition to a more local neighbourhood-oriented service for a diverse public of users. The question then becomes to what extent the knowledges and infrastructures that are currently being accumulated and constructed in the Romanian church can be supported by the state and gradually be opened up to a more diverse public.

In a way, this is exactly the transition the unionized railway employees in my research have gone through. For decades, their colleagues were white Belgians with long-term contracts. The solidary practices of the organization have sedimented in strong vertical connections and rigid procedures. Yet, over the past decade, these local union members have faced increasing pressure to dismantle the strong railway employee statute and have found themselves 'thrown together' (Massey, 2005) with a diversifying group of colleagues who carry the burden of cost-cutting strategies: they have lower salaries, different contracts, speak different tongues and have different meritocratic moralities.

While these new colleagues were initially seen as foreign competitors, prolonged exposure has helped to reframe them as victims of the same exploitative mechanisms the union had been criticizing for some time. The density of union force made it possible to connect different unions (the railway union and the cleaners' union). The strong vertical organization of the union helped to stage the victimhood of the Romanians quickly in the national press. Throughout this process, the Romanians learned to think of themselves as colleagues with potential organizational power and not only as individuals who have to organize their own welfare. Hence, it can be seen that 'throwntogetherness' does a lot in activating, negotiating and renewing pre-existing mechanisms of solidarity.

Drawing on the conceptualization of relational places developed by Amin (2004) and Massey (2005), the question is whether the new situational we-ness – and the solidarities it sustains – is limited to very specific places in the arrival infrastructure of Brussels or whether it can also expand to other places. In the vocabulary of Massey (2005), the 'outwardlookingness' of these solidary initiatives in Brussels remains unclear. What aspects of the pedagogical environment of Brussels described in this chapter feed back into the broader politics of moral transformation in Romania outlined at the start of this chapter? Will it be the stories of individual success that further endorse the myth of the 'successful returning migrant entrepreneur' who is best suited to lead the post-socialist future? Will it be stories of faith- and ethnicity-based solidarity that crosses class divides and inspires charity-based forms of solidarity in Romania? Or will it be the experience of solidarity in diversity and the difficult but not impossible revitalization of strong solidarity infrastructures such as trade unions?

Notes

1 The narrative of the poor, but entrepreneurial migrant also resonates with the lingering growth of a particular strand of liberal urban governance discourses on 'slum' and 'ethnic' entrepreneurship, such as Saunders's (2010) account of the 'arrival city'. As revealed by McFarlane (2012) and Amin (2013), these discourses narrow down the energies in poor immigrant urban areas to the entrepreneurial potentials of individuals and do not scale up the innovative practices of – and discourses on – solidarity-in-diversity that grow there as well in a similar way (Oosterlynck et al., 2016).

2 Attracting more than 100,000 consumers, the Abattoir market is the largest open-air market in Belgium. It sells cheap food and second-hand materials and hosts a number of shops and pubs selling Romanian food and drinks. On lampposts and walls, a myriad of little papers in Romanian inform customers about different work, housing, transport and leisure opportunities.

3 The situation is completely different in Flanders, where the number of NMS migrants employed with a permit B is much higher as a result of the common use of the permit for seasonal work in agriculture (Wets & Pauwels, 2011).

4 With a work permit B, a system segmenting the Belgian labour market first established in 1936 at a moment of economic crisis, non-Belgian workers are allowed to work only for a single Belgian employer and for a fixed period. Hence, in principle, the permit B does not allow labour market mobility. However, the permit B can be extended four

times after which a permit A can be obtained (Caestecker, 2000). Another important form of employment is the posted worker. The number of Romanian posted workers registered in the Limosa system more than doubled between 2008 (3,263) and 2012 (8,318) in the whole of Belgium (Mussche et al., 2013).

5 All names of institutions and respondents have been anonymized.

6 The significance of the concentration of diverse syndical forces in the train station cannot be underestimated. According to one of my respondents, the NMBS-holding, probably aware of this danger, has recently started moving the more 'risky' outsourced contracts to workplaces with a lower syndical density such as the hangars.

References

Amin, A. (2004). Regions unbound: Towards a new politics of place. *Geografiska Annaler: Series B: Human Geography, 86*(1), 33–44.

Amin, A. (2013). Telescopic urbanism and the poor. *City, 17*(4), 476–492.

Badescu, G. (2004). Romanian labor migration and citizenship. In D. Pop (Ed.), *New patterns of labor migration in Central and Eastern Europe* (pp. 8–18). Cluj Napoca: Amm Press.

Blommaert, J. (2011). The Vatican of the diaspora. *Jaarboek voor Lithurgie-onderzoek, 27*, 243–259.

Bourgeois, G. (2009). *Beleidsnota Inburgering & Integratie 2009–2014* [Policy paper on citizenship making & integration 2009–2014]. Retrieved from www.inburgering. be/inburgering/sites/www.inburgering.be.inburgering/files/Beleidsnota_Inburgering_ Integratie_2009-2014.pdf.

Butler, J. (1988). Performative acts and gender constitution: An essay in phenomenology and feminist theory. *Theatre Journal, 40*(4), 519–531.

Caestecker, F. (2000). *Alien policy in Belgium, 1840–1940: The creation of guest workers, refugees and illegal aliens*. New York and Oxford: Berghahn Books.

Castles, S., & Kosack, G. (1973). *Immigrant workers and class structure in Western Europe*. London: Oxford University Press.

Danckaers, T. (2012, 22 February). België blijft Roemenen en Bulgaren weren [Belgium keeps refusing Romanians and Bulgarians]. *Mo* Mondiaal Nieuws*, retrieved from www.mo.be/en/node/6217.

Datta, A. (2009). 'This is special humour': Visual narratives of Polish masculinities in London's building sites. In K. Burrell (Ed.), *After 2004: Polish migration to the UK in the 'New' European Union* (pp. 189–210). Aldershot: Ashgate.

Datta, A., & Brickell, K. (2009). 'We have a little bit more finesse, as a nation': Constructing the Polish worker in London's building sites. *Antipode, 41*(3), 439–464.

Digneffe, J. (2008). *ACOD-spoor schenkt klare wijn rond syndicale werkingsmiddelen* [ACOD-rail clarifies its operating resources]. Retrieved from www.acod-spoor.be/index. php?option=com_content&view=article&id=274:acod-spoor-schenkt-bklare-wijn-rond-syndicale-werkingsmidellen&catid=57:laatste-nieuws (last accessed 26 January 2016).

Fox, J. E., & Jones, D. (2013). Migration, everyday life and the ethnicity bias. *Ethnicities, 13*(4), 385–400.

Fox, J. E., Moroşanu, L., & Szilassy, E. (2015). Denying discrimination: Status, 'race', and the whitening of Britain's New Europeans. *Journal of Ethnic and Migration Studies, 41*(5), 729–748.

Glick Schiller, N. (2008). *Beyond methodological ethnicity: Local and transnational pathways of immigrant incorporation.* Willy Brandt Series of Working Papers in International Migration and Ethnic Relations 2/08. Malmö: MIM and IMER.

Guldentops, F. (2012). *NMBS-SNCB. Stop sociale dumping* [Stop social dumping]. Retrieved from www.petities24.com/nmbs-sncb_stop_sociale_dumping_-_stop_au_dumping_social (last accessed 26 February 2016).

Guldentops, F. (2014, 25 January). De schoonmakers in de schaduw van de NMBS-stations [The janitors in the shadow of the railway stations]. Interview with Dominique Fervaille, *De Nieuwe Socialist*. Retrieved from https://denieuwesocialist.wordpress.com/2014/01/25/de-schoonmakers-in-de-schaduw-van-de-nmbs-stations-interview-met-dominique-fervaille/ (last accessed 28 June 2016).

Hermia, J. P. (2015). Een demografische boom onder de loep: Roemenen, Polen en Bulgaren in het Brussels Hoofdstedelijk Gewest [A demographic boom under scrutiny: Romanians, Poles and Bulgarians in the Brussels Capital Region]. *BISA Focus*, *9*, 1–9.

Kilkey, M., & Perrons, D. (2010). Gendered divisions in domestic work time: The rise of the (migrant) handyman phenomenon. *Time & Society*, *19*(2), 239–264.

Kloosterman, R., van der Leun, J. P., & Rath, J. (1999). Mixed embeddedness: Immigrant entrepreneurship and informal economic activities. *International Journal of Urban and Regional Research*, *23*(2), 253–267.

Krings, T. (2009). A race to the bottom? Trade unions, EU enlargement and the free movement of labour. *European Journal of Industrial Relations*, *15*(1), 49–69.

Levin, I. (2014). 'This is for the children, the grandchildren. . .': Houses of Moroccan immigrants in metropolitan Tel Aviv. *Housing, Theory and Society*, *31*(1), 54–75.

Ley, D. (2008). The immigrant church as an urban service hub. *Urban Studies*, *45*(10), 2057–2074.

Massey, D. (2005). *For space*. London: SAGE.

McFarlane, C. (2012). The entrepreneurial slum: Civil society, mobility and the co-production of urban development. *Urban Studies*, *49*(13), 2795–2816.

McKay, S. C. (2007). Filipino sea men: Constructing masculinities in an ethnic labour niche. *Journal of Ethnic and Migration Studies*, *33*(4), 617–633.

Meeus, B. (2013). Welfare through migrant work: What if the Romanian 'safety valve' closes? *Southeast European and Black Sea Studies*, *13*(2), 175–194.

Meeus, B., & Schillebeeckx, E. (2015). Geloofsgeïnspireerde organisaties en de woonnood van nieuwkomers in stedelijke aankomstwijken [Faith-based organizations and the housing needs of newcomers in urban neighbourhoods of arrival]. In P. De Decker, B. Meeus, I. Pannecoucke, E. Schillebeeckx, J. Verstraete & E. Volckaert (Eds.), *Woonnood in Vlaanderen. Feiten-Mythen-Voorstellen* [*Housing needs in Flanders. Facts-myths-proposals*] (pp. 495–516). Antwerpen-Apeldoorn: Garant.

Meeus, B., van Heur, B., & Arnaut, K. (forthcoming). *Arrival infrastructures*. London: Palgrave Macmillan.

Moriarty, E., Wickham, J., Bobek, A., & Daly, S. (2015). Portability of social protection in the European Union: A transformation of national welfare systems? In A. Amelina, K. Horvath & B. Meeus (Eds.), *Anthology of migration and social transformation: European perspectives* (pp. 201–215). Heidelberg: Springer.

Moroşanu, L., & Fox, J. E. (2013). 'No smoke without fire': Strategies of coping with stigmatized migrant identities. *Ethnicities*, *13*(4), 438–456.

Mullings, B. (2012). Governmentality, diaspora assemblages and the ongoing challenge of 'development'. *Antipode*, *44*(2), 406–427.

Mussche, N., Corluy, V., Marx, I., & Haemels, J. (2013). *Arbeidsmarktonderzoek als instrument en basis bij toekomstig arbeidsmigratiebeleid en EU vrijhandelsakkoorden.* [Labour market research as tool and basis for future labour market policy and EU free trade agreements]. Antwerpen: Universiteit Antwerpen.

Oosterlynck, S., Loopmans, M., Schuermans, N., Vandenabeele, J., & Zemni, S. (2016). Putting flesh to the bone: Looking for solidarity in diversity, here and now. *Ethnic and Racial Studies*, *39*(5), 764–782.

Parutis, V. (2011). White, European, and hardworking: East European migrants' relationships with other communities in London. *Journal of Baltic Studies*, *42*(2), 263–288.

Paspalanova, M. (2006). *Undocumented and legal Eastern European immigrants in Brussels.* Unpublished doctoral dissertation, KU Leuven.

Pijpers, R. (2006). Help! The Poles are coming: Narrating a contemporary moral panic. *Geografiska Annaler B*, *88*(1), 91–103.

Portes, A. (1995). *The economic sociology of immigration. Essays on networks, ethnicity and entrepreneurship.* New York: Russell Sage Foundation.

Portes, A., & Sensenbrenner, J. (1993). Embeddedness and immigration: Notes on the social determinants of economic action. *The American Journal of Sociology*, *98*(6), 1320–1350.

Rath, J. (2002). A quintessential immigrant niche? The non-case of immigrants in the Dutch construction industry. *Entrepreneurship & Regional Development*, *14*, 355–372.

Rath, J., & Swagerman, A. (2011). *Promoting ethnic entrepreneurship in European cities.* Luxembourg: Publications Office of the European Union.

Saunders, D. (2010). *Arrival city: How the largest migration in history is reshaping our world.* London: William Heinemann.

Schrover, M., Van der Leun, J., & Quispel, C. (2007). Niches, labour market segregation, ethnicity and gender. *Journal of Ethnic and Migration Studies*, *33*(4), 529–540.

Soenens, D. (2012, 24 February). De vuile hoekjes van de schoonmaak [The dirty corners of the cleaning business]. *Vacature*. Retrieved from www.vacature.com/artikel/de-vuile-hoekjes-van-de-schoonmaak (last accessed 26 January 2016).

Sullivan, T. (2012). 'I want to be all I can Irish': The role of performance and performativity in the construction of ethnicity. *Social & Cultural Geography*, *13*(5), 429–443.

Tamas, K., & Münz, R. (2006). *Labour migrants unbound? EU enlargement, transitional measures and labour market effects.* Stockholm: Institute for Futures Studies.

Trappers, A. (2009). *Relations, reputations, regulations: An anthropological study of the integration of Romanian immigrants in Brussels, Lisbon and Stockholm.* Unpublished doctoral dissertation, KU Leuven.

Vandaele, K., & Faniel, J. (2012). Geen grenzen aan de groei: de Belgische syndicalisatiegraad in de jaren 2000 [No limits to the growth; the Belgian degree of unionization in the 2000s]. *Over.Werk*, *22*(4), 124–132.

Waldinger, R. (1995). The 'other side' of embeddedness: A case-study of the interplay of economy and ethnicity. *Ethnic and Racial Studies*, *18*(3), 555–580.

Wets, J., & Pauwels, F. (2011). *Arbeidsmigratie vanuit Oost-Europa. Polen, Roemenen en Bulgaren op de Belgische arbeidsmarkt* [Labour migration from Eastern Europe. Poles, Romanians and Bulgarians on the Belgian labour market]. Leuven: HIVA.

Wills, J. (2008). Making class politics possible: Organizing contract cleaners in London. *International Journal of Urban and Regional Research*, *32*(2), 305–323.

7 The spatial solidarity of intentional neighbouring

Andy Walter, Katherine Hankins and Samuel Nowak

Introduction

In 2005, at the age of 25, Christy Norwood, who by all appearances was a 'white suburban girl', moved to the inner-city neighbourhood of South Atlanta, just south of Atlanta's downtown core. At the time, South Atlanta was over 90 per cent African American with over 30 per cent of the households living below the poverty line. It was rather unusual for a college-educated, white girl to sign the lease papers for a house on a street where crack and prostitution houses were not uncommon. Christy moved to the inner city, as have thousands of other middle-class people of faith in cities and struggling suburbs across the United States, to live as an 'intentional neighbour'. In what follows, we examine the ways in which intentional neighbouring demonstrates the enactment of faith-based solidarity rooted in encounter. In doing so, we highlight the importance of *spatial solidarity* as both a guiding logic and a socio-spatial practice that emerges from the sustained experience of the daily, weekly and yearly encounters of intentional neighbours who live their lives in the marginalized spaces of the city.

The 'three-Rs' strategy that underpins Christian community development was articulated by an African American pastor from Mississippi named John Perkins. In the 1970s, he became frustrated by the Civil Rights Movement's loss of focus on improving the concrete conditions of people's lives or, as he put it, 'making beloved community happen. . . in the places where we are' (Perkins, 2009; see Hankins et al., 2015). Perkins was also deeply disgruntled with the Christian church, which too readily accommodated itself to an unjust economic system and racist state institutions. As he reflects in *With justice for all*, the gospel has been 'robbed. . . of its power' to challenge structures of oppression and exploitation (Perkins, 2007, p. 30). For Perkins, disrupting the unjust and exclusionary status quo required a shift away from serving poor and racially segregated communities at a distance to living in propinquity through a new model of Christian community development.

To that end, during the 1970s and 1980s Perkins worked out his 'three Rs' strategy. The 'R' of reconciliation explicitly entails efforts to break down social boundaries enabling relations of oppression and exclusion between groups across lines of class, and especially race. Reconciliation is 'a strategy for here and now', an embodied commitment to living with difference (Perkins, 2007). It is not achieved through 'social mix' or mere diversity in a place but through attempting to build mutually respectful relationships across racial and cultural differences. Reconciliation inevitably requires addressing exploitative economic relations, as accumulated inequalities in wealth and material resources present significant barriers between and among social groups. Achieving reconciliation, thus, involves a second 'R', standing for redistribution, calling on 'community developers' to confront the processes and structures by which resources and opportunities are socially allocated and to work to alter them to the benefit of historically exploited groups. The third 'R' represents 'relocation'. Christian community developers must 'move outside the privileged space to the place of marginality' (Gornik, 2002, p. 31) in order to develop embodied relationships with fellow, if different, human beings and to discern their problems and aspirations in concrete, contextual terms. As Perkins argued, '[r]elocation is the method by which we accomplish reconciliation and redistribution. Neither reconciliation nor redistribution can be done effectively long distance. . . Relocation is personal' (Perkins, 1996, p. 36). For, '[w]ithout relocation, without living among the people. . . it is impossible to accurately identify the needs as the people perceive them' (Perkins, 2007, p. 65).

For Christy Norwood, and thousands of other intentional neighbours across the United States, the 'three Rs' are the faithful enactment of a socio-spatial logic that entails a commitment to sustained encounter with difference for purposes of racial reconciliation and economic redistribution. The spatial solidarity of intentional neighbouring, then, is based on the faith-motivated call for Christians to live in close propinquity to the poor and racially segregated in order to know them and their needs. This knowing, as we demonstrate below, takes place through encounters, or interactions across social difference. Intentional neighbours foster these deliberate interactions with their neighbours in order to establish the basis for solidarity with them. They engage in repeated encounters in everyday spaces, such as the home or sidewalk, to develop the embodied relationships that allow them to understand more fully the lived experience of the poor.

Solidarity, a concept that is receiving increasing attention in social sciences, as this volume attests, suggests a sharing of values and common responsibility within and across social groups and 'an attitude of compassionate reciprocity, aimed at achieving a social order that ensures mutual respect and a decent life for all' (Nel & Taylor, 2013, p. 1091). In much social science literature, solidarity has been connected to social groups within or across territorial nation-states, and as Oosterlynck et al. (2016, p. 765) rightly point out, this focus misses the opportunity to understand the 'solidarities that develop in the different spatio-temporal register of everyday place-based practices'. Neighbourhoods and the everyday interactions that take place within them provide one such setting from which to examine emergent or potential solidarities of the 'here and now'. We see the possibility of solidarity through the faith-based principles of intentional neighbouring, which requires people of faith to move into high-poverty places and engage in interdependent relationships with neighbours, learn their values and share the struggles they face in their daily lives. In this chapter, we focus specifically on intentional neighbours' encounters with difference in order to explore in depth the role of faith in developing place-based solidarities. While solidarity is necessarily relational, we focus, in this chapter, on the potential for solidarity to emerge among intentional neighbours and their poor and racialized neighbours against structures of oppression.

In what follows, we explore the writings and philosophies of Christian community development (CCD) and draw out the spatial solidarity that Perkins and other CCD architects require of themselves and others. We then explore the experience of encounter among 25 intentional neighbours who have been living in (or have recently lived in) high-poverty neighbourhoods in Atlanta, Georgia. Most of our informants are white middle-class men and women associated with FCS Urban Ministries, a large Christian community development organization that is grounded in Perkins's and Robert Lupton's ideas and has operated in the city for more than 35 years (Lupton, 1997, 2005, 2007, 2011, 2012). Some of these intentional neighbours were part of the Mission Year program, an FCS affiliate program that situates young men and women in high-poverty neighbourhoods in select American cities. We focus on the ways in which their spatial solidarity creates opportunities for meaningful contact with social difference, highlighting this solidarity as a guiding logic for faith-based neighbouring.

The spatial solidarity of Christian community development

Intentional neighbouring is part of a larger tide of Christian community development. Across the United States and through international networks on multiple continents, thousands of people are putting Perkins's ideas into practice. Perkins has published a number of widely read books, is a sought-after public speaker, and in 1989 he joined with like-minded activists from around the United States to form the Christian Community Development Association (CCDA), an organization whose core philosophy is the 'three Rs'. Today, the CCDA is a nationwide, multi-racial/ethnic network of more than 3,000 individual members and nearly 400 partner organizations. Numerous widely read authors and practitioners of Christian community development are associated with the organization, including Robert Lupton working in Atlanta (Lupton, 1997, 2005, 2007, 2011, 2012; see Hankins & Walter, 2012); Ray Bakke in Chicago and then Seattle (Bakke, 1997); Mary Nelson and Wayne Gordon in Chicago (Nelson, 2010; Gordon, 1995); Mark Gornik and Alan Tibbels in Baltimore (Gornik, 2002); and Shane Claiborne in Philadelphia (Claiborne, 2006), among others.

The 'R' of relocation reveals a place-based concept of solidarity at the heart of Christian community development praxis. Across a range of writers (e.g. Perkins, Gornik, Lupton) and in CCDA promotional materials and practitioner guides, Christians are encouraged to act on the biblical call to 'do justice' by relocating themselves in order to develop interpersonal relationships with the poor, provide practical support for them, and accept some degree of accountability for their well-being (harkening to the etymological origin of the term solidarity from Roman law, *obligato in solidum*, which refers to the common liability of citizens for a debt; Nel & Taylor, 2013, p. 1091). The practice of 'neighbouring' is not just a declaration of love for fellow human beings but a demonstration of 'brotherhood' with particular ones, specifically those who are socially powerless, marginalized, excluded and exploited. In doing so, intentional neighbours are understood to be following the model of Jesus. As Perkins explains, 'Jesus relocated. He didn't commute to Earth one day a week and shoot back up to heaven. He left his throne and became one of us. . . The incarnation is the ultimate relocation' (Perkins, 2007, p. 90). Relocation is not only an expression of faith, however, but is also grounded in a social analysis revealing to Perkins and these other writers the necessity of propinquity in achieving the goals of Christian community development.

In Christian community development writings, place and propinquity are linked to the effective pursuit of justice for poor and marginalized people. These texts express a contextualized understanding of poverty that resists narratives of personal responsibility and moral failure, as well as recognizing the uneven geographies of access to opportunities, resources and rights (Gornik, 2002; Claiborne, 2006). For these writers, being in solidarity with the poor necessarily involves place-focused work, or place-making (see Hankins & Walter, 2012). CCD practice is also informed by a concern with distance, in particular the kind of distance associated with spatial marginality, where the poor are isolated in neglected pockets of the city. This socio-spatial distance contributes to gaps in understanding what the concrete challenges are for poor people in their places. For many policymakers, diagnosing what is 'needed' in high-poverty places generally occurs from the 'outside' of those places. The spatial propinquity of intentional neighbouring is not the only way to understand the conditions of the poor, nor does it guarantee understanding, but for Perkins and others, it is a necessary approach. Acting at a geographical distance, or 'from the outside' (Perkins, 2007, p. 92) tends to result in patronizing, predetermined and ineffective, if not harmful, practices. Achieving propinquitous relationships to others allows 'relocaters' to listen and learn about the concrete challenges people face, enabling practical support and greater possibilities to close the social, emotional and cognitive distance, ultimately paving the way for solidarities to emerge.

Our contribution to the growing literature on solidarity is three-fold. The first is empirical: we contribute to a larger project of working to understand intentional neighbouring, which has been largely overlooked in geography and other social science research despite its inherent focus on place and social justice. With some notable exceptions, such as Beaumont (2008a, 2008b, 2008c), Beaumont and Baker (2011), Cloke (2011) and others (e.g. Sziarto, 2008; Warren, 2008), geographers have been relatively silent on faith-motivated practices that open up social and spatial justice possibilities.

Second, we explicitly name *spatial solidarity* as a socio-spatial practice. By spatial solidarity we refer to the place-based concerns and shared values that emerge from the daily interactions of propinquitous living. As we discuss below, we recognize that not all propinquitous living results in such solidarity, as much of the social mix literature has demonstrated (e.g. Lees, 2008; Davidson, 2010), but the example of intentional neighbouring provides a novel empirical case that exposes the possibility for solidarity in diversity.

Third, following Oosterlynck et al. (2016), we examine encounter as a source of solidarity, where we emphasize the moments in the everyday spaces of the home or the sidewalk, where intentional neighbours seek to form relationships across various axes of difference which impacts their own socio-spatial positionality (Shepherd, 2002) and subjectivity. By developing relationships with other residents of their neighbourhood, by learning about their living situations, their families, their employment histories and their aspirations, they learn of the conditions that shape the way residents experience poverty and marginalization and how their own privilege (whether middle-classness and/or whiteness) is implicated in that marginalization. In so doing, their encounters become a potential source of solidarity among diversity.

Encounters with difference are key in intentional neighbouring. The act of neighbouring inevitably involves contact and interactions with residents who are often marginalized as poor, racialized Others. The 'contact hypothesis', emerging from social psychology (Allport, 1954), suggests that under certain conditions, contact between majority and minority groups can reduce negative attitudes of the former towards the latter. Geographers have suggested that spatial proximity, however, is not enough to guarantee such meaningful or transformative contact (Valentine, 2008; Leitner, 2012). At the same time, however, 'the encounter holds open the possibility of not only inscribing but also disorienting us from the habits, stereotypes, and prejudices toward the Other, creating the possibility for change and transformation' (Leitner, 2012, p. 820).

To date, the encounter literature has focused primarily on fleeting encounters in public space (Valentine, 2008; Wilson, 2011) or on what Amin (2002) calls micro-publics: spaces such as community centres (Matejskova & Leitner, 2011) or participatory art projects (Askins & Pain, 2011); neighbourhood planning meetings (Lawson & Elwood, 2013) or the intimate spaces of the home (Schuermans, 2013). A growing area of research within this literature has focused on the process of neighbouring – sustained contact and place-making in the space of the neighbourhood – through, for instance, social mix housing policies and gentrification (Lees, 2008; Davidson, 2010). In particular, research has examined encounters between white middle-class residents and poor and racialized others, concluding that middle-class groups, while eager consumers of cultural diversity, frequently do not engage in meaningful interactions with their neighbours (Lees, 2008; Pinkster, 2014). Rather, they often consolidate their privilege, reproducing middle-class norms (Elwood et al., 2015) and creating socio-spatial enclaves that limit encounter with original residents (van Eijk, 2010; Pinkster, 2014). Other research suggests that diversity is not an important factor motivating middle-class actors to move into marginalized, heterogeneous neighbourhoods, and that affordable housing prices and proximity to the downtown core are the primary factors (Albeda et al., 2015).

In this chapter, we suggest that intentional neighbours represent rather different white middle-class subjects than commonly studied in the literature on neighbourhood encounters. Intentional neighbours do not relocate to poor neighbourhoods in the 'normal' course of their middle-class lives. They are not there for affordability or accessibility, nor for facile consumption of diversity. Rather, they are primarily motivated by their faith to live in spatial solidarity with the marginalized. It is their faith that provides the ontological and epistemological basis for relocating, and therefore it is the basis from which encounters with social difference in the neighbourhood originate. As one intentional neighbour Sarah says so succinctly, 'we live there on purpose. We didn't just happen upon this community.' For these neighbours, passing encounters with social difference are not merely a by-product of their choice to relocate; rather intentional engagement is the very reason for their presence in the neighbourhood. It is in this sense that they appear to seek what Valentine (2008, p. 325) calls 'meaningful contact', social interaction that 'actually changes values and translates into. . . positive respect for – rather than merely a tolerance of – others'. In what follows, we use a narrow definition of Valentine's concept of 'meaningful contact' – that is, it entails a change in socio-spatial positionality on the part of the more privileged party, not necessarily a shared, reciprocal trust. While intentional neighbours often speak about such 'mutual respect', our empirics limit our ability to adequately address the mutuality of these statements. Instead, we focus in on the processes of subjectification engendered by intentional neighbours' encounters with socio-spatial difference, their attempts to make 'meaningful contact', and the socio-spatial experience that guides their praxis – the spatial solidarity of intentional neighbouring.

Intentional neighbouring in Atlanta, Georgia

In our study, which was conducted between 2010 and 2013, we interviewed 16 intentional neighbours who were associated with the Mission Year program (see Dahl, 2012). Many of these young men and women lived in either English Avenue, which is located northwest of downtown Atlanta, or Polar Rock, which is situated directly south of downtown. In addition, we interviewed nine intentional neighbours residing in South Atlanta, a neighbourhood just slightly north and east of Polar Rock. All of these neighbourhoods are majority African American with high rates of vacant and abandoned homes and with poverty rates exceeding 30 per cent, according to the US Census. We interviewed eight men and 17 women, ranging in age from 20 years old to those in their early 40s. Twenty-one were white, two identified as Latina and two as Asian. These intentional neighbours moved into these neighbourhoods from a range of places, such as from other areas of Atlanta, from the suburbs of Lansing, Michigan and from the inner city of Toronto, Canada, to name just a few. At the time of our interviews, the intentional neighbours living in South Atlanta had been living there anywhere from one year to over a decade.

In what follows, we identify three dimensions of the spatial solidarity of intentional neighbouring. First, intentional neighbours seek to know and experience the material conditions of marginalized neighbourhoods. They deliberately relocate from spaces of relative privilege in order to experience the spatial context of poverty. Relocation, however, is a necessary but not sufficient condition for spatial solidarity, as the social mix literature has consistently demonstrated (e.g. Lees, 2008). Thus, there is a second dimension of spatial solidarity: intentional neighbours seek out, and create conditions for, encounters with their neighbours. They do not insulate themselves from their neighbours, but rather intentionally spend time on their porches and in communal spaces, they share their food, offer rides and create conditions for encounter through everyday processes of place-making. Third, encounter does not necessarily translate into 'meaningful contact' (Valentine, 2008). We highlight, therefore, how intentional neighbours shift their understandings of poverty and privilege through everyday agonisms. We show how encounters with social difference can translate into meaningful contact on the part of intentional neighbours. This tripartite is both the logic and socio-spatial practice that guides the spatial solidarity of intentional neighbouring.

Solidarity and place

A key element of developing solidarity between intentional neighbours and 'already there' residents is in developing a deep understanding of the context in which the poor and marginalized live. For many of our interview participants, the place of the neighbourhood itself shaped their understanding of their neighbours and created possibilities for further solidarities to emerge from their shared experience of place. Quite clearly, intentional neighbours sought out particular places that they perceived as in need of resources, but the very lack of those resources that they experienced on a daily basis was transformative for many intentional neighbours we interviewed.

Mission Year participant Kelly recounts her experience of the English Avenue neighbourhood and the residents she had gotten to know there:

> My experience with homelessness before has been on the streets of downtown Toronto. It's similar to downtown Atlanta where you see people living on the street, but I've never seen urban poverty where it's almost like suffering in your own home. You have a shelter, a roof over your head, but you don't own it or can't afford to pay rent. Or you have a roof over your head but you can't afford dinner or the commodities of living in that home. That was a shock. And even to see the state that homes are in. Houses are falling apart, there's abandonment.

Kelly shares that the condition of much of the housing in English Avenue is shocking to her and the lived experience of poverty of many of the neighbourhood's residents is profoundly different from anything she knew growing up (see Figures 7.1 and 7.2). Through her exposure to the daily context in which her neighbours live, she experiences a shift in her understanding of poverty from just 'the visible poor' (Blau, 1993), to a much more nuanced appreciation of the conditions of the working poor.

Figure 7.1 Abandoned house near the English Avenue Mission Year house. Photograph by Traci Dahl (2012).

Figure 7.2 Vacant apartment building near the English Avenue Mission Year house. Photograph by Traci Dahl (2012).

Another Mission Year participant Ann describes the lack of services in her neighbourhood, such as public bus stops and library facilities. She rhetorically asks:

> How come some parts of the city, particularly where we are living, they don't have great MARTA [public transit] service? It's pretty inaccessible. We don't have resources. Like, the closest library to our house, I can't get there.

How's that fair? Who is in charge of the library? . . . [To get to the library] you have to take the bus out of our subdivision to the North Avenue Station and take the train to [the Five Points Station], then get on another bus and essentially come back, like do a big U. Why? Why can't they have bus routes in our part of the city? There's no grocery stores, either [*sic*].

For Ann, where she lives is *her* part of the city. It is where she lives and experiences the materiality of the neighbourhood, which makes her aware of the unfairness of the differential access to public services and retail outlets that she and her neighbours experience in English Avenue. Through her experience of the place, she shares in the struggles associated with being a resident of a high-poverty, under-resourced neighbourhood; it becomes her struggle.

The Mission Year participants ended their program after one year, and many chose to remain in the neighbourhood where they had been living. Others moved back home, went on to finish college, or moved into an adjacent neighbourhood to continue intentional neighbouring. The choice to leave is one that differentiates intentional neighbours from many other residents in the neighbourhood, but their experience of living in a high-poverty neighbourhood gives them an understanding of the context for how their socially marginalized neighbours live. South Atlanta intentional neighbour Christy reflects on this:

I still experience some of the same injustices [as my neighbours] and so I can actually understand better [the challenges they face], but I still have to remember: I can leave. And that does make a difference. But at least I know. The light goes out on a beautiful day, I can experience that frustration. Or the police not coming to 911 calls, but I could move somewhere where they would respond to my 911 call.

This recognition that intentional neighbours have a choice to leave high-poverty neighbourhoods and rejoin middle-class landscapes makes plain to them the profound lack of choices their working-class and poorer neighbours have. But the fact that many of them experience one year or choose to stay for years or even decades reveals to them the material and experiential knowledge of marginalized neighbourhoods.

Not only is the space of the neighbourhood a source of potential solidarity, but as we suggest, encounter between intentional neighbours and 'already there' residents provides an important source of understanding the experience of poverty and marginalization. In fact, it is explicitly what Perkins (2007, p. 65) suggests is required of Christian community developers: to know 'the needs as the people perceive them'.

Encountering neighbours

Lynn, a South Atlanta resident, describes her efforts at meeting her neighbours on the street where she and her husband rent their home:

We've tried to be real intentional from the beginning to go over and introduce ourselves, and try to be outside [of our house]. . . You know, we've tried to intentionalize things like Mother's Day. You know, we've taken planters of flowers over to – there are, both my neighbours, they're single moms – you know, I can tell that that really impacted them, just different things. Birthdays, I've offered to make cakes for my neighbours, and they're all the time like, 'Yeah, we would love that'.

Likewise, many of the Mission Year participants recounted their experiences of inviting their neighbours over to share the limited amount of food they had (Mission Year restricts the weekly house budget of its participants). One former Mission Year participant Shelly recalled that she and her housemates 'would have, oh my gosh, ridiculous things like two boxes of macaroni and cheese and a bowl full of pretzels, and then maybe some kind of old apples, just. . . ridiculous things that we would feed people'. She reflected that 'it. . . just. . . became a way of life, like just feeding people what you have versus making a big ordeal about it. If you have it, you give it, but if you don't, you don't.' The point for Shelly and Lynn and for other intentional neighbours we interviewed was to create conditions to develop meaningful contact with their neighbours, to attempt to form relationships with them in order to know them.

In addition to deliberate, everyday encounters, intentional neighbours see that part of their goal is to have sustained engagements with their neighbours over long periods of time. As Dan, an intentional neighbour of South Atlanta, suggests, time is a key component to developing meaningful relationships:

It takes a little bit more time and intentionality to connect with people [who] are from the neighbourhood [compared to other intentional neighbours]. And we have really good relationships, but again, it just takes time. It takes a lot of intentional time because we have to say no to certain things to say yes to other things, right? We have a lot of friends who live in the suburbs and a lot of family that lives in the suburbs, and you know, if we said yes to that, to connecting with them, we'd be saying no to hanging out in our neighbourhood. Like, this past July 4th, we said no to hanging out with family and friends in other places. Saying, 'no we're going to share life here'. . . and it was great.

Dan makes the point that investing in developing propinquitous relationships takes a sustained commitment to a place that can mean rejecting opportunities to be elsewhere. In order to understand the concrete problems and aspirations of his neighbours, he has to be present, in their shared spaces, to interact with them. Intentional neighbours, by their very name and purpose, do not 'just live here' (Pinkster, 2014), they make conscious choices to share life with their neighbours. Rather than remaining in networked spaces of relative privilege and isolating themselves from the neighbourhood, they seek to connect with it and to learn the values and struggles of their neighbours.

In some cases, intentional neighbours sought to create places where sustained, repetitive encounters could occur. For example, one intentional neighbour in South Atlanta is also the owner of the neighbourhood's coffee shop, an enterprise that is part of the FCS Urban Ministries economic development initiative. Jeff reflects on the role of the coffee shop in providing an important space of encounter for neighbourhood residents:

> We believe that there's a value of having a neighbourhood gathering place that people can come to and, you know, hang out. We say we're not gonna kick 'em out if they don't buy anything. There's no hourly limit. There's no library close by. There's no public access to Internet, so we want a place where adults can look for a job [and] students can come down and do their homework. . . from a business sense, especially from a place where there's not a lot of disposable income, that doesn't help your business very much, but it does fit into why we're here.

He goes on to say, 'if our only goal was to make money, we would have closed up a long time ago'. For Jeff, the coffee shop provides a space for encounters, where neighbours can have normalized, everyday interactions. He expresses that he values the space for creating the possibility for these encounters and not merely for financial gain.

As Perkins and other CCD writers emphasize, the goal of relocating is to establish relationships based on mutual respect and reciprocity. This depends on earning the trust of others, requiring intentional neighbours to demonstrate humility and acceptance of people for who they are and to practice patience, understanding people's suspicion of themselves as outsiders. For example, Clint, an intentional neighbour in South Atlanta reflects on what might be termed breakthrough encounters after he had been living in the neighbourhood for a while:

> The thing about moving into any [neighbourhood]. . . You know, you really gotta acclimate to it. . . We're not from Atlanta, not from this neighbourhood. It just takes time for that [trust] to be built. At a time it was a big deal for our neighbours to just invite us into their home just for a little bit, not even like we're gonna come in and talk, just like I need to borrow something. Instead of me waiting outside, they're like 'oh, you can come in'. . . that's somewhat of a victory.

For Clint, being invited into the house suggested for him the possibility of some level of trust from his neighbours. The development of trust in neighbourly relationships was sought after by many of our participants. For some intentional neighbours, then, sustaining the deliberate encounters, such as taking the Mother's Day floral baskets to neighbours, led to what many describe as moments when they felt or experienced what they perceived as trust.

Everyday efforts to connect with their neighbours, however, are not always rewarded with the trust they seek. Understandably, many 'already there' residents are sceptical of, or even outright hostile towards, intentional neighbours. Coping with this seemingly insurmountable sense of distrust and sustained failure to connect with their neighbours led many intentional neighbours to feelings of what they term 'burn out'. Case (2011, p. 84) explores the frustrations expressed by the 32 intentional neighbours she interviewed, concluding that '[d]espite their ideals and "long-term" commitment to "stay", most participants shared that they had thoughts about moving out or leaving, specifically when things "got hard" and they felt overwhelmed with the high degree of need within the neighbourhood'. 'Burn out' can lead intentional neighbours to question not only their commitment to neighbouring, but also their religious beliefs, causing some to leave the neighbourhood altogether after experiencing 'crises of faith' (ibid., p. 111). In this sense, intentional neighbouring is hard emotional labour that requires sometimes more than they are capable or willing to give or to sustain indefinitely, suggesting that there is, for some, a limit to the spatial solidarity of intentional neighbouring.

Solidarity and privilege

For many of our participants, the encounters they experience in their intentional neighbouring practices shape how they came to understand the contours of social and spatial injustice and the experience of poverty. Many of them made personal connections with their neighbours and subsequently developed a sense of the structural and social pressures that shape their neighbours' daily lives. This has led many of them to reject the narratives of personal responsibility and individual failings that dominate neoliberal accounts of poverty (and that shape anti-poverty strategies in the neoliberal city). As Camarin puts it:

> I definitely [thought] the most oppressed people [was] the middle class right now, because we don't have any privileges. Like everyone has all of these laws that benefit them, and I'm like, 'I don't have any laws that benefit me'. . . [Living in South Atlanta] I definitely explored it even further and understand it even more concretely and more completely, the idea of white privilege and racism and classism and sexism and kind of the intertwining of all of those and how. . . I didn't deal with [them] growing up. It wasn't in my face. As a white male, middle class, I have an option. . . And that is privilege.

Before relocating to the inner city, Camarin saw the middle class as oppressed by an unfair system that heaps benefits and protective laws on the poor and marginalized, while the middle-class enjoy no such advantages. He saw privilege as endowed by the state in the form of social welfare and civil rights, not as the product of racist legacies, spatial processes of accumulation by dispossession or advantages conferred by identity. Valentine (2008, p. 333) maintains that such a victimhood mentality severely limits the potential for encounters to break down prejudice: '[t]he reason that such individual everyday encounters do not necessarily change people's general prejudices is because they do not destabilise white majority community based narratives of economic and/or cultural victim-hood'. What we see from Camarin, and in the case of intentional neighbours more broadly, appears to confirm this claim. After relocating to the neighbourhood, Camarin and other intentional neighbours cannot resort to easy and unquestioned narratives of victimization. They encounter the very real injustices of oppression in their everyday life and in doing so actively and constantly engage with their privilege, even naming structures of racism, classism, sexism that were not apparent to them when they lived in more homogenous, middle-class spaces.

In this process of knowing their neighbours and living in high-poverty places, intentional neighbours begin to engage with social difference on different terms, challenging dominant narratives of poverty and privilege. As Mission Year participant Ann states,

> [What living in a high poverty neighbourhood] has done is given me an opportunity for people back home. . . . I'm telling people that I'm working with homeless and low income people and they're, like, 'Oh it's their fault' and I'm like, 'Well actually, no, let me tell you about it. Let me tell how my neighbours are doing everything they can. Let me tell you about the person who eats at the soup kitchen who has a family and a degree from the country where he came from but can't get a green card here.' And stuff like that. It's given me an opportunity to share that it's not all poor decisions, there's people who are trying really hard, and it could happen to any of us.

Ann sums it up by saying, 'I've met people who have told me their stories and they are doing everything they can but they can't catch a break.'

An awareness of the structures that shape the ability of residents of high-poverty neighbourhoods 'to catch a break' is also apparent in Christy's experience in getting to know several of the long-time residents of South Atlanta. She connects their individual stories with the historical legacies of systemic disinvestment and discrimination that characterizes the treatment of African Americans in the US:

> And so to understand some of the poverty that exists now. . . How can you expect a good system, how can you expect good reading levels when we set up a shitty school system here [in this city]? And you had slavery. You had segregation. I mean Miss Mary Porter [a long-time resident of South Atlanta] was talking about how they didn't have books [in schools for African American children] and all this stuff. . . . A lot of social justice issues here can be connected to that. . . . You know, as a white person, I have something to say about that. I need – there is some part of rectifying that, and it's not that long ago, though we want to think it was long ago, but if I can talk to someone who lived through it, it's not.

For Christy, the personal connection of knowing someone who can describe the overcrowded schools without books that African American children attended is connected to the current conditions of poverty. By knowing her neighbour Miss Porter, by developing a relationship with her, Christy learns stories about Miss Porter's youth, about growing up as an African American child in the segregated South. The history and legacy of institutionalized racism becomes clear for Christy, who may not have fully appreciated its dynamics had she lived in a typical middle-class neighbourhood or failed to reach out and make meaningful contact with her neighbours.

For Mission Year participant Jenny, getting to know the mothers in her neighbourhood revealed to her the structural conditions that make providing for their children difficult:

> My idea [in helping in my neighbourhood] was that I could love little kids who don't have moms who love them, ya know. But I didn't think about. . . when you're in the situation, that's when your eyes are opened to the fact that life is not fair. It is not necessarily. . . this mom's fault that she can't provide for her child, but it's because of the system that we have in our government, in our society that is prohibiting her from getting a job where she can provide for her child.

By making personal connections, intentional neighbours like Jenny begin to more fully understand the broader social conditions that shape the experience of living in poverty in the twenty-first-century city. She learns that it is not the fault of the mother who juggles parenting and working, but rather the insufficiency of the social safety net in the United States, where childcare costs are prohibitively high, welfare benefits are extremely limited and minimum wage does not provide a liveable wage. Living with the poor enables intentional neighbours to know their neighbours and to develop multiple sources of solidarity (Oosterlynck et al., 2016) around a shared set of values, such as a strong family life or the struggle against inadequate wages. It is in this sense that the subjectivities of intentional neighbours shift from a prescriptive attitude that assumes they are the normative model against which their neighbours should be compared, to a 'willingness to negotiate the diversity of people and the practices that they are engaged in here and now' (Oosterlynck et al., 2016, p. 775).

The personal relationships through sustained, everyday encounters underlie the solidarity that intentional neighbours seek to develop. And it is the relationships that for many intentional neighbours maintain focus on various forms and expressions of injustice. Sarah, an intentional neighbour in South Atlanta married a South American immigrant and found herself engaged in broader political battles around immigration. She recounts:

> I think the first couple of years that I started doing this I kept [my interactions] very one-to-one. And I remember being asked. . . what's the role of. . . [political] advocacy? And I was, like, oh, you know, that's for radicals or whatever. And I think probably partly because of immigration [issues], you couldn't keep it on a one-on-one level. Like immigration can't be dealt with

between me and you. Like you have to have some sweeping changes. And so what does that mean to kind of leverage your privilege and stand in that gap to be able to say, like, 'These are my neighbours, these are my friends and some of them literally cannot stand up for themselves without severe risk so I'm going to have to get involved on their behalf.' And so, yeah, now I don't keep it as much to the one-on-one, but I think it's a good starting place. And it's something. . . if I spent too much time just thinking about immigration from, like, a policy and advocacy standpoint, I would just jump over a cliff I think. And so. . . the reality is. . . immigrants are having birthday parties and are going to work and are celebrating things. And so having those relationships also sustains the broader work, I think. . . That's where you get. That's where life happens and the relationships, too.

For Sarah, the structural conditions, such as immigrant rights, forces her to contend with larger, nation-state policies. And she feels strongly that because of her middle-class, educated position as a US citizen, she can stand up for an undocumented immigrant. She sees that she can use her privilege for advocacy in a way that her marginalized neighbours cannot. It is through daily connections to people's lives, through learning their stories, through deliberate or breakthrough daily encounters, through *knowing* that her advocacy is sustained and her solidarity with her neighbours can emerge.

Conclusions

At the outset of this chapter, we suggested that intentional neighbouring provides an opening to consider the contours of solidarity in diversity. The guiding logic of Christian community development with its 'three Rs' strategy involving relocation, redistribution and racial reconciliation reveals that faith can be an important motivation for different forms of solidarity around social and spatial justice. Driven by the theologies of John Perkins, intentional neighbours place themselves in areas of socio-spatial marginality, deliberately removing themselves from the networked spaces of relative privilege to which they are accustomed. Often for the first time, they experience what it is like to have unpaved or pothole-stricken roads, to live next to vacant houses, to live in a food desert, to experience the frustration when the city's bus service is cut, when it takes police hours to respond and when one's neighbourhood is a known dumping ground for trash and tires. Intentional neighbours learn intimately about the spatial context of where the poor live their lives. In contrast to middle-class subjects commonly studied in the literature on neighbouring, they do not insulate themselves from encounters with social difference, but instead seek them out, reworking both their own understandings of poverty and privilege and neighbourhood-level networks. As much literature in geography and sociology has pointed out, neither residential propinquity, nor encounters with social difference, equate to meaningful interactions in the spaces of the neighbourhood, but taken together with the faith-based motivation of understanding the needs of the poor and marginalized, we see that encounter initiated and sustained by intentional neighbours has the potential to shape their own middle-class subjectivities, revealing structural forces of oppression that their neighbours experience.

Spatial solidarity, as we name it, involves developing common understandings and collective responsibility around place-based experiences, concerns and struggles. Intentional neighbours are there to work towards racial reconciliation and economic redistribution – and to 'destabilize[e] boundaries and creat[e] new spaces for negotiating across difference' (Leitner, 2012, p. 830). As we demonstrate, intentional neighbours create these new spaces for engaging difference and destabilizing their own understandings of poverty and privilege, upsetting pre-existing narratives of middle-class victimhood that impede 'meaningful contact'. This is not to say their attempts at meaningful contact are always successful, nor that we can take their claims of mutual trust at face value, given the notable silence of 'already there' residents in this chapter. What we can conclude, however, is that intentional neighbours deploy a faith-based spatial solidarity that, while rooted in everyday encounter, operates at multiple spatial registers. Their relocation, and attempts at redistribution and reconciliation, rework neighbourhood-level networks and relationships, individual place-making practices and political commitments. As Sarah's case suggests, meaningful contact in the neighbourhood motivates her advocacy to work for immigrant rights at broader spatial scales. In this sense, the spatial solidarity of intentional neighbouring shapes and reinforces other sources of solidarity such as interdependence, shared values and common struggles that could lead to collective action and new forms of living together in difference.

References

Albeda, Y., Oosterlynck, S., Verschraegen, G., Saeys, A. & Dierckx, D. (2015). *Governing urban diversity: Creating social cohesion, social mobility and economic performance in today's hyper-diversified cities report 2b fieldwork inhabitants, Antwerp (Belgium)*. Retrieved from www.urbandivercities.eu/wp-content/uploads/2015/08/Belgium_WP6_final_report.pdf.

Allport, G. (1954). *The nature of prejudice*. Reading, MA: Addison-Wesley.

Amin, A. (2002). Ethnicity and the multicultural city: Living with diversity. *Environment and Planning A, 34*(6), 959–980.

Askins, K., & Pain, R. (2011). Contact zones: Participation, materiality, and the messiness of interaction. *Environment and Planning D: Society and Space, 29*, 803–821.

Bakke, R. (1997). *A theology as big as the city*. Downers Grove, IL: InterVarsity Press.

Beaumont, J. (2008a). Introduction: Dossier on faith-based organizations and human geography. *Tijdschrift voor Economische en Sociale Geografie, 99*, 377–381.

Beaumont, J. (2008b). Introduction: Faith-based organizations and urban social issues. *Urban Studies, 45*, 2011–2017.

Beaumont, J. (2008c). Faith action on urban social issues. *Urban Studies, 45*, 2019–2034.

Beaumont, J., & Baker, C. (Eds.). (2011). *Postsecular cities: Religious space, theory and practice*. London and New York: Continuum International.

Blau, J. (1993). *The visible poor: Homelessness in the United States*. New York: Oxford University Press.

Case, C. (2011). *Strategic neighboring and 'beloved community' development in west Atlanta neighborhoods.* Unpublished Master's thesis, Georgia State University, Atlanta.

Claiborne, S. (2006). *The irresistible revolution: Living as an ordinary radical.* Grand Rapids, MI: Zondervan Publishing House.

Cloke, P. (2011). Geography and invisible powers. In D. Harvey, N. Thomas, C. Brace, A. Bailey & S. Carter (Eds.), *Emerging geographies of belief* (pp. 9–29). Newcastle upon Tyne: Cambridge Scholars.

Dahl, T. (2012). *Shifting conceptions of social justice in faith-based care workers as a result of the Mission Year Program.* Unpublished Master's thesis, Georgia State University, Atlanta.

Davidson, M. (2010). Love thy neighbor? Social mixing in London's gentrification frontiers. *Environment and Planning A, 42*(3), 524–544.

Elwood, S., Lawson, V., & Nowak, S. (2015). Middle-class poverty politics: Making place, making people. *Annals of the Association of American Geographers, 105*(1), 123–143.

Gordon, W. (1995). *Real hope in Chicago.* Grand Rapids, MI: Zondervan Publishing House.

Gornik, M. (2002). *To live in peace: Biblical faith and the changing inner city.* Grand Rapids, MI and Cambridge, UK: William B. Eerdmans Publishing.

Hankins, K., & Walter, A. (2012). 'Gentrification with justice': An urban ministry collective and the practice of place-making in Atlanta's inner city neighborhoods. *Urban Studies, 49*(7), 1507–1526.

Hankins, K., Walter, A., & Derickson, K. (2015). 'Committing to a place': The place-based, faith-based legacies of the Civil Rights Movement. *Political Geography, 48*, 159–168.

Lawson, V., & Elwood, S. (2013). Encountering poverty: Space, class, and poverty politics. *Antipode, 46*(1), 209–228.

Lees, L. (2008). Gentrification and social mixing: Towards an inclusive urban renaissance? *Urban Studies, 45*(12), 2449–2470.

Leitner, H. (2012). Spaces of encounter: Immigration, race, class, and the politics of belonging in small-town America. *Annals of the Association of American Geographers, 102*(4), 828–846.

Lupton, R. (1997). *Return flight: Community development through reneighboring our cities.* Atlanta, GA: FCS Urban Ministries.

Lupton, R. (2005). *Renewing the city: Reflections on community development and urban renewal.* Downers Grove, IL: InterVarsity Press.

Lupton, R. (2007). *Compassion, justice, and the Christian life: Rethinking ministry to the poor.* Grand Rapids, MI: Baker Books.

Lupton, R. (2011). *Theirs is the kingdom: Celebrating the Gospel in urban America.* New York: HarperCollins.

Lupton, R. (2012). *Toxic charity: How churches and charities hurt those they help, and how to reverse it.* New York: HarperCollins.

Matejskova, T., & Leitner, H. (2011). Urban encounters with difference: The contact hypothesis and immigrant integration projects in eastern Berlin. *Social and Cultural Geography, 12*(7), 717–741.

Nel, P., & Taylor, I. (2013). Bugger thy neighbour? IBSA and south-south solidarity. *Third World Quarterly, 34*(6), 1091–1110.

Nelson, M. (2010). *Empowerment: A key component of Christian community development.* Christian Community Development Association. Bloomington, IN: iUniverse.

Oosterlynck, S., Loopmans, M., Schuermans, N., Vandenabeele, J., & Zemni, S. (2016). Putting flesh to the bone: Looking for solidarity in diversity, here and now. *Ethnic and Racial Studies, 39*, 764–782.

Perkins, J. (Ed.). (1996). *Restoring at-risk communities: Doing it together and doing it right*. Grand Rapids, MI: Baker Books.

Perkins, J. (2007). *With justice for all: A strategy for community development*. Venture, CA: Regal.

Perkins, J. (2009). A time for rebuilding. In C. Marsh & J. Perkins (Eds.), *Welcoming justice* (pp. 107–120). Downer's Grove, IL: IVP Books.

Pinkster, F. M. (2014). 'I just live here': Everyday practices of disaffiliation of middle-class households in disadvantaged neighbourhoods. *Urban Studies, 51*(4), 810–826.

Schuermans, N. (2013). Ambivalent geographies of encounter inside and around the fortified homes of middle class Whites in Cape Town. *Journal of Housing and the Built Environment, 28*, 679–688.

Shepherd, E. (2002). The spaces and times of globalization: Place, scale, networks and positionality. *Economic Geography, 78*(3), 307–330.

Sziarto, K. (2008). Placing legitimacy: Organizing religious support in a hospital workers' contract campaign. *Tijdschrift voor Economische en Sociale Geografie, 99*(4), 400–425.

Valentine, G. (2008). Living with difference: Reflections on geographies of encounter. *Progress in Human Geography, 32*(3), 323–337.

van Eijk, G. (2010). *Unequal networks spatial segregation, relationships and inequality in the city*. Amsterdam: IOS Press.

Warren, M. (2008). A theology of organizing: From Alinsky to the modern IAF. In J. Defilippis & S. Saegert (Eds.), *The community development reader* (pp. 194–203). New York and London: Routledge.

Wilson, H. F. (2011). Passing propinquities in the multicultural city: The everyday encounters of bus passengering. *Environment and Planning A, 43*(3), 634–649.

8 Football for solidarity

Bridging gaps between the Baka and the Bantu in East Cameroon

Harrison Esam Awuh and Floor Elisabeth Spijkers

Introduction

Speculation about effects of intergroup contact dates back to the nineteenth century and studies on the topic started as early as the 1930s (Pettigrew & Tropp, 2005). In this field, Allport's (1954) work has shown to be most influential. Ever since its formulation, his contact hypothesis has inspired hundreds of studies (Pettigrew et al., 2011). In a meta-analysis of more than 500 of these studies, Pettigrew and Tropp (2006) concluded that the hypothesis is generally supported and that the conditions set by Allport (equal status, institutional supports, common interests and the pursuit of common goals) turn out to be facilitating, indeed. As such, it is worthwhile to quote Allport (1954, p. 281) here at length:

> Prejudice (unless deeply rooted in the character structure of the individual) may be reduced by equal status contact between majority and minority groups in the pursuit of common goals. The effect is greatly enhanced if this contact is sanctioned by institutional supports (by law, custom or local atmosphere), and provided it is of a sort that leads to the perception of common interests and common humanity between members of the two groups.

Obviously, it is essential to distinguish between different types of contact. According to the mere exposure effect (Zajonc, 1968), repeated exposure to an object or person already increases our liking of that object or person (e.g. Moreland & Beach, 1992; Burger et al., 2001; Jones et al., 2011). Other scholars argue, however, that just seeing each other is not enough and claim that such superficial forms of contact are likely to trigger stereotypes and negative associations and can, therefore, lead to an increase of negative out-group sentiments (Barlow et al., 2012; Schuermans, 2016). As such, they underline the importance of deeper intergroup contact (Amin, 2002; Pettigrew, 1998). Drawing on a case study in the Marzahn neighbourhood of Berlin, Matejskova and Leitner (2011) infer, for instance, that fundamental changes in the capacity to live with difference are generally not achieved in fleeting exchanges in public spaces, but in regular and shared activities. Based on 18 months of ethnographic fieldwork in the London borough of Hackney, Wessendorf (2013) confirms that regular encounters in social spaces provided by community organizations or parental groups in primary schools enhance a deeper intercultural understanding, as they increase communication and, hence, familiarity with different others.

Drawing on a football project in East Cameroon, this chapter aims to contribute to contact research in at least two ways. First, we find it necessary to reflect upon the effects of contact. Most research on the contact hypothesis looks at shifts in attitudes. By focusing on what people think and say, and by neglecting what people do, much scholarship in this field cannot rule out the possibility that there is an inconsistency between the values expressed in survey research and the practices people actually deploy towards others (Valentine, 2008). To overcome this gap between values and practices, we will focus, in this chapter, on an increase of solidarity, rather than a reduction of prejudices. In the definition of Stjernø (2004, p. 25), solidarity complements an attitudinal component (feelings of shared fate and group loyalty) with a behavioural one (the sharing and redistribution of resources such as time and money). Similarly, for Bayertz (1999, p. 3), solidarity does not only encompass 'a factual level of actual common ground between individuals', but also 'a normative level of mutual obligations to aid each other, as and when should be necessary'.

Apart from an analysis of the effects of contact, this chapter will also focus on the conditions under which the contact takes place. According to the original hypothesis proposed by Allport (1954), a couple of conditions need to be met in order for contact to reduce prejudices. As has been indicated above, these conditions revolve around the equal status of the participants, the pursuit of common goals, the cooperation between different groups, the perception of a common humanity and institutional support. Our study will investigate the importance of these conditions, but add an additional one. Inspired by the literature on the geographies of encounter (e.g. Valentine, 2008; Matejskova & Leitner, 2011; Schuermans, 2016), we will make it clear that the place where contact occurs is not neutral and has an important effect on the effects of contact.

To substantiate these two claims, this chapter presents the results of a football programme in Cameroon which was set up as an action research project to improve the relationships between the local Bantu and Baka communities. By stimulating positive intergroup interactions, sports activities have often been considered to be avenues for social integration, social cohesion, social capital and social solidarity (De Knop & Elling, 2000; Verweel et al., 2005; Ratna, 2010; Theeboom et al., 2012; Spracklen, 2013). Suggesting that sport could be used to promote social inclusion, the UK government responded to racial unrest in towns in northern England, for instance, by developing a number of cross-department Policy Action Teams (Long et al., 2009). Yet, research also shows that tensions in inter-ethnic sports contexts can inhibit the participation of certain ethnic groups (Janssens, 1999; Van der Meulen, 2007) and that intergroup competition may also create strong outgroup antagonism (Elling, 2004; Coffé & Geys, 2006). As a result, Hoffmann (2002) acknowledges that sports activities are not integrative per se, but that the conditions under which they take place determine their integrative potential.

As such, this chapter will focus on the conditions under which a football tournament set up as an action research project in Cameroon was successful in nurturing solidarities. The next two sections will introduce our quantitative and qualitative research methodologies and the specific research context of the villages around the Dja Reserve in East Cameroon. Afterwards, the first empirical section will focus on the effects of contact during the football tournament. Before coming to the conclusion, the second empirical section will focus on the conditions under which such contact has positive effects. In the latter section, special attention will be paid to the place of contact.

Setting the scene

Our study is based in the villages of Bingongol, Mintoum, Adjela, Abakoum, Le Bosquet and Payo. They are located in East Cameroon around the Dja Reserve (Figure 8.1). In this area, one finds representatives of both Baka and Bantu ethnic groups. Traditionally, the Bantu are sedentary farmers. The Nzime make up the main Bantu group in the area. Around the mid-nineteenth century, they migrated south because of conflicts with the Beti who were themselves pushed south by the Vute and Mbum people fleeing Fulani warriors (Ngoh, 1996). In the literature on the central African sub-region, the Baka are conventionally known as hunter-gatherers. They are considered to be the first inhabitants of the Dja Reserve area (Bahuchet, 2000). Historically, Baka groups settled in small groups and led a semi-nomadic life based on a widespread knowledge of the natural resources in their area. Baka culture and religion were also strongly linked to the forest (Shikongo, 2005).

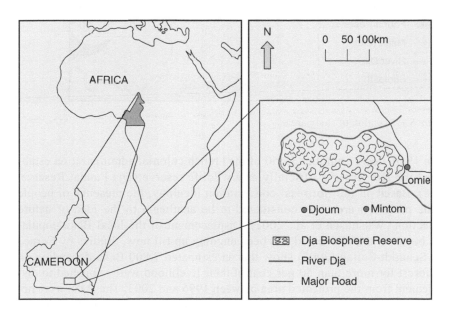

Figure 8.1 (continued)

(continued)

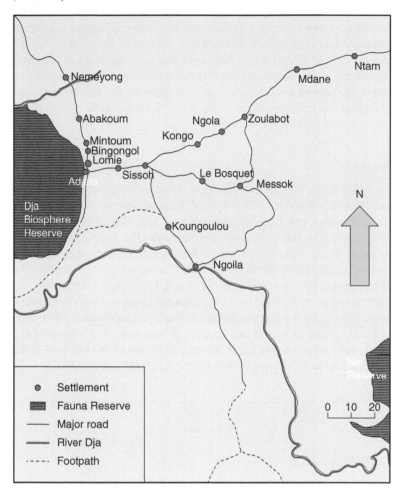

Figure 8.1 Location of study area.

In 1950, decree number 75/50 of the French colonial administration established the Dja Reserve officially as a wildlife reserve (Dja Faunal Reserve,
2008). Based on the 'fortress' conservation ideology, the presence of people
in the protected area was considered to be anathema to the idea of nature
protection (Wilshusen et al., 2002). Displacement of the local Baka population began in the 1950s and has been ongoing up till now. Studies by Cernea
and Schmidt-Soltau (2003) show that an estimated 7,800 Baka depending on
the forest for more than 50 per cent of their livelihood were subjected to displacement from the protected area between 1996 and 2003. The Baka were not
consulted prior to displacement. No compensation was offered for their loss of
land and livelihood either.

Together with the disruption of the traditional lifestyle, traditional forms of solidarity eroded. Several socio-cultural changes related to conservation-induced displacement and resettlement are responsible for this (Awuh, 2016). A first threat to traditional forms of solidarity in the Baka community is that traditional modes of exchange based on reciprocity and redistribution have been replaced by a cash economy. According to Bahuchet (1990), redistribution in Baka society involves collective food-getting and food-sharing activities. During gathering, several people – both women (for plants or insects) and men (for honey) – work together and share the food (Bahuchet, 1985). During hunting, meat is divided among hunters. Afterwards, each hunter shares the meat with his extended family. Cooked food is also distributed within the family (Bahuchet, 1990). While most forms of redistribution comprise horizontal social networks of kinship and friendship, a successful hunter would, in some instances, also distribute his kill over several people with the expectation of a return gift to make up for the debt (*ekola*) (Bahuchet, 1990).

Following displacement, reciprocity has come under increasing threat as a mode of exchange. The Bantu economy operates principally in cash, so Baka working for Bantu are introduced to the cash economy. Eighty per cent of the resettled Baka have switched to cash as a mode of exchange for labour and 83 per cent use cash as a mode of exchange for goods and other services (Awuh, 2016). Even among the Baka themselves, preference for reciprocity has diminished. For instance, up to half of all transactions among the Baka in Adjela are currently based on cash (Awuh, 2016). The introduction of the cash economy puts pressure on the classic forms of solidarity in the Baka community. With exchanges depending on amount of cash to spend, inequalities within the Baka community also became substantial. Not all Baka have paid jobs. Those who have often earn much less than their Bantu colleagues.

Traditional forms of solidarity among the Baka are also threatened by decreased involvement in community tasks and activities. Conventionally, the Baka lived in very tightly knit communities. However, following displacement, participation in group activities with other Baka people has diminished (Awuh, 2016). This is evident in the *Jengi* ritual. The Baka worship a forest spirit referred to as *Jengi*. *Jengi* is believed to be omnipresent within the forest and has the powers to punish transgressors and protect the faithful from the dangers of the forest. Besides being the spirit of the forest, the traditional initiation ceremony is also referred to as *Jengi*. Traditionally, every Baka man attends *Jengi* about three to four times a year. Such collective rituals strengthen feelings of community and enhance solidarity (Fischer et al., 2013). However, following displacement, one third of Baka adults in Adjela have never been initiated into *Jengi*. In Le Bosquet, the figure amounts to 40 per cent. These figures are high for varied reasons. In Adjela the high figure was blamed on men being more involved in working for the Bantu (which gives them less time to be involved with the Baka community), whereas in Le Bosquet Christianity was seen as the major threat to *Jengi*.

Crucially, the loss of solidarity in the Baka community has not been replaced with solidarity between the Baka and the Bantu. Although Baka are in contact with the Bantu through work and living side by side, most contact remains shallow and functional. Baka settlements are often administratively considered as part of the neighbouring Bantu settlements. For instance, the Baka settlement of Bingongol is considered part of the neighbouring Bantu settlement with the same name. Although the Baka and the Bantu live in the same village officially, they practically live within their own communities and they hold negative stereotypes towards each other. In previous interviews with Baka and Bantu participants, the Baka considered the Bantu as arrogant, selfish and untrustworthy. The Bantu, on the other hand, regard the Baka as unreliable, dirty, alcoholic, short-sighted, irrational and illiterate (Awuh, 2016). Not only did the Baka lose their livelihood following displacement, they are, thus, also facing marginalization. In any case, relations between Baka and Bantu in the villages around the Dja Reserve have shown to be tense.

Methods

In this chapter, we argue that better relations between the Baka and the Bantu are crucial. As stereotyping and lack of positive interactions between Baka and Bantu are identified as crucial barriers to intergroup solidarity, an action research intervention was set up in which Baka and Bantu were brought together in positive contact situations. The intervention consisted of two main actions: a football event and digging of latrines in Baka villages. With the action research, we aimed to take a first step in the development of intergroup solidarity between Baka and Bantu community members.

In order to do justice to the contextual rootedness and specificity of social relations over time and space (see Brewer & Miller, 1984), we allowed participants to determine whether they wanted to meet with the other group and, if so, which activities they would like to pursue. The Baka participants noted that Baka and Bantu often live parallel lives and that they would like to have a football tournament as an opportunity to interact more with Bantu community members. As the football event was part of a larger project concerning the improvement of Baka livelihood, the Baka participants also envisioned using sports as a means to construct latrines in the villages.

Each of the six teams participating in the tournament was composed of 11 players, a coach and a coordinator. Even though the majority of participants came from the Baka community, each team consisted of representatives of both communities. The research team randomly selected 30 Bantu from the different villages for participation in the programme. Questionnaires were administered before the tournament to identity where these Bantu held friendship ties with Baka or not. Twenty-five out of the 30 Bantu men had no friendship ties with any Baka people. All players were male and aged between 16 and 40 years. In each village, committees were set up to coordinate the football programme. In most cases, the coordinator was an elderly or influential member of the community. One person per committee was also trained as a football coach. Over a three-week period, teams organized about five training sessions per week lasting an average of about two hours per day. The football tournament itself was a one-day event (see Figures 8.2 and 8.3). The referees at the football tournament were carefully

Figure 8.2 Images from the weekend of games in Lomie town.

Figure 8.3 Baka and Bantu playing together (top) and celebrating together (bottom).

chosen by the research team. In the context of tense relations between Baka and Bantu, it was decided upon to select referees who did not belong to either of the local communities. At all stages of the football programme, participants were reminded that their participation was voluntary and they should reserve the right to withdraw.

In a preceding photovoice exercise (Awuh, 2016), the Baka identified poor sanitation and particularly the lack of latrines as a major community concern following displacement. Prior to the football programme, only eight latrines existed across the six Baka villages under study (approximately 400 to 1,500 people per village). As much as 84 per cent of the Baka did not have access to any latrines or toilets. The absence of sanitation infrastructure had a strong impact on the health of the inhabitants of these villages. According to Dounias and Froment (2006), high rates of intestinal worms are a direct consequence of increased contact with all sorts of faecal waste. Such contacts cause intestinal infections of bacterial and viral origins and are the principal cause of infectious diarrhoea, anaemia and delayed growth among children, with potentially dramatic consequences for their psychological development.

As such, the Baka decided upon the construction of a minimum number of latrines (five per village and ten in Le Bosquet because of its relatively larger population size) as the main requirement for participation in the football competition. The latrine construction was done per football team. Baka and Bantu of the same teams constructed latrines in the villages of the respective Baka players (see Figure 8.4). Although participation in the latrine digging process was prerequisite to participation in the football programme for the Baka,[1] the Bantu participants were under no obligation to help the Baka.[2] Nonetheless, even though the Bantu players were invited to voluntarily participate in the exercise of digging latrines in Baka communities, all Bantu players decided to help the Baka with the construction of latrines. Latrine digging was done principally in the morning period over a three-week period. The evaluation of latrine digging was done two days prior to the football competition by the research team[3] and the committees. Despite all attempts to get the men to fully participate in the project, the village of Adjela was unanimously excluded from the competition due to the failure of its inhabitants to construct any latrines (Figure 8.5). In the other villages, a total of 28 new latrines was constructed.

As quantitative and qualitative methodologies each have their own strengths and weaknesses, both structured questionnaires and in-depth interviews were used to analyse the effects of contact during the football competition and latrine digging. Considering the target population is predominantly illiterate, structured questionnaires were administered by the research team to 40 Baka and 20 Bantu who participated in the football programme. To identify the effect of the initiative, this happened both before and after the intervention. The questionnaires addressed interethnic relations between Baka and Bantu. Items included: 'Would you marry a Baka/Bantu?', 'Would you let your son or daughter marry a Baka/Bantu?', 'Do you have any Bantu/Baka friends?' and 'Would you make friends with Bantu/Baka people?' Answers to the questions were categorized as

Figure 8.4 Dug latrines in Le Bosquet and Mintoum.

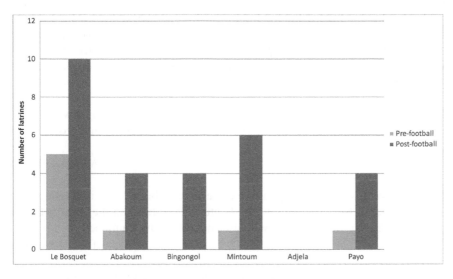

Figure 8.5 Football and latrine construction in six villages.

'Yes', 'No' or 'Not sure'. The order of questions on the post-football programme questionnaires were reversed and mixed with regular or 'dummy' demographic questions in order to diminish respondent bias from fatigue, predictable question order and social desirability.

As closed-ended questions fail to adequately explore the meanings participants attach to the contact (Cherry, 1995; Dixon et al., 2005), 12 additional in-depth semi-structured interviews (four with Bantu and eight with Baka) were conducted to facilitate the interpretation of trends identified in the questionnaires. Each of the semi-structured interviews lasted between 30 to 45 minutes. Generally, participants were questioned on intergroup attitudes and intergroup contact. Open-ended questions allowed participants to expand on their meanings, interpretations and experiences of contact with the other group before and after the intervention. In order to minimize response bias, both the interviews and the questionnaires were carried out at least three days after the event itself. By introducing a time lag between the measurement of the effect of the football programme and the actual football event, 'mood of the moment' responses were minimized (see Podsakoff et al., 2003).

Stereotypes, discourses and practices

Most respondents in our study regarded sport as an important contributor to intergroup contact. According to our survey, 85 per cent of the Baka and 80 per cent percent of the Bantu respondents viewed football as an important medium for social contact with the other ethnic group. Contact through sport also changed the stereotypes Baka and Bantu had about each other. While two out of 20 Bantu respondents did not change their attitudes towards the Baka after the programme, the remaining 18 acknowledged that they had done so. Baka respondents indicated that they held more positive views about the Bantu after the football programme as well. In the interviews with open-ended questions, the importance of the intervention for improving relations between Baka and Bantu was emphasized by participants from both communities:[4]

> I think it was a good platform to meet and interact with people we will not normally interact with. It was great to have fun with our Bantu brothers and forge friendships with them.
>
> (M, Baka player)

> I believe this football programme was a success because by interacting with the Baka in this competition, the Bantu will understand the Baka better in times of crisis and the Baka will also understand the Bantu better.
>
> (L, Baka coach)

In the quantitative survey, a considerable number of respondents indicated that they had changed their views about marrying a member of the opposite ethnic group after the football programme. Before the football programme, 24 out of 40 Baka affirmed they could marry a Bantu person, while only eight out of 20 Bantu participants declared they could marry a Baka person. In contrast, after the football programme, all of the 40 Baka respondents affirmed they could marry a Bantu person, whereas 14 out of 20 Bantu confirmed they could marry a Baka person. Likewise, before the football programme, 75 per cent of the Baka and 60 per cent of the Bantu participants asserted they could permit their offspring to marry a member of the other group. After the football programme, these percentages amounted to 90 and 70 per cent, respectively. While the marginalized Baka are more willing to surrender prejudice because marrying or letting an offspring marry a Bantu could be a symbol of prestige, the Bantu view things differently. For the Bantu, marrying a Baka means that they marry someone with less status. In addition, the Bantu look down at the hunter-gatherer lifestyle of the Baka, which they consider to be primitive. This could explain the smaller number of Bantu willing to intermarry with the Baka in spite of the progress made by joint participation in this football programme. A Bantu player affirms that:

> I can make friends with the Baka but I cannot marry a Baka woman. Our lifestyles are too different[5] and that is why I think a marriage will not work. At the moment I cannot say if I will let my offspring marry a Baka or not because it will be the personal choice of my son or daughter.
>
> (JC, Bantu player)

Importantly, the football tournament did not only have effects on the attitudes and the discourses of our respondents, but also upon their practices. After the football tournament, many Bantu indicated that they would treat the Baka differently, and vice versa. Drawing on the positive contact they had experienced during the training sessions, the matches and the digging of latrines, several respondents from both ethnic groups stated that they were more eager to develop relationships of reciprocity with members of the other group. In the latrine digging exercise, the Bantu helped the Baka to construct latrines in their villages without getting anything back. As such, the action research project we set up did not only impact upon the attitudes of the Bantu and the Baka, but also upon their willingness to foster reciprocal relationships of solidarity outside the cash economy:

> It is amazing to see how the Baka with assistance from the Bantu were will-ing to dig their own latrines without any financial aid whatsoever. I believe this will teach us all to take more responsibility for changes in our communi-ties without external help.
>
> (F, Bantu player)

> Normally, the Baka are reluctant to help the Bantu for free. However, in this programme I saw the Bantu helping in the construction of latrines for the Baka for nothing. I think this sends a message to both communities that we can help each other without always expecting a financial reward.
>
> (AP, Baka player)

In some instances, Bantu participants even indicated that they would cancel the debts of Baka working for them:

> Personally, even if a Baka in my team owed me money before the football competition, after moments of so much fun together on the football field as we had in this football competition, I will be tempted to cancel the debt.
>
> (AL, Bantu player)

> There was a Baka in the team we played our first match against who did some work for me the day after the football final. I paid him 3,000Frs[6] for the work he did. However, he owed a Bantu 5,000Frs.[7] This Bantu heard from other people that I had paid the Baka some money. So, he went over with a group of other Bantus and seized the Baka who could not pay all the money owed. When I was told about this, I went over and confronted the Bantu in person. We argued and I decided to pay all the money the Baka owed him in order to secure the release of the Baka. This happened just yesterday. I can assure you that I could not have done that for a Baka before this football programme. This has really changed me in the way I see the Baka.
>
> (S, Bantu player)

All in all, these quotations demonstrate that the friendly interactions during the football tournament, the preparatory trainings and the digging of latrines helped both the Baka and the Bantu to reconsider the way they deal with each other. While contact between both groups is usually based on unequal relationships of exchange, the football project allowed the participants to change the way they think about each other, and hence, to establish more reciprocal relationships based on trust, mutual help, responsibility and solidarity. To understand the success of the football tournament in changing the stereotypes and the practices of the participants, the next section will return to Allport's (1954) set of conditions under which contact reduces prejudices.

A place-based perspective

As indicated in the introduction of this chapter, a first criterion proposed by Allport is that both groups have an equal status in the contact situation. By creating mixed football teams of Bantu and Baka, we met this condition. While differences between the Baka and the Bantu are explicit in day-to-day life, this project emphasized their communalities. Usually, the Baka work for the Bantu, who claim a superior status. In the football project, however, members of both groups dug latrines side by side, played football matches together and trained as a group. In the tournament, Baka and Bantu were teammates and football lovers and this enabled bonding. Because Baka and Bantu encountered each other on an equal basis – as 'brothers', as it is called in the quote below – they could develop a spirit of solidarity based on equality:

> In this football competition, there were no fights as is often the case here with football. We played together as brothers with the Baka. This shows that there are no major differences between the Baka and the Bantu.
>
> (P, Bantu player)

A second condition proposed by Allport revolves around the pursuit of common goals. In the case of our action research intervention, these goals were set by the research team, in close cooperation with the participants themselves. When asked why they wanted to participate in the football programme, more than half of the Baka respondents gave 'networking with Bantu' as their main motivation. And while only one Bantu respondent viewed networking with the Baka as the main driver behind his involvement, more than half of the Bantu cited the need for community development as their main reason for participating in the programme. Yet, for most participants, the ultimate goal was to win the football tournament. In order to hold the trophy in the air, they had to train together and dig latrines. Hence, football served as a reward system for participation in community work which in turn generated deeper contact between the Baka and the Bantu. In this way, the pursuit of different common goals led to increased solidarity:

> I think trust has definitely improved. I will not say every Baka in the football team will trust every Bantu now because of the football competition. However, I think the spirit of being together and working together for common goals makes a significant difference.
>
> (M, Bantu player)

Normally, the Baka are wary of the Bantu in spite of the fact that we are human beings like them. By interacting with the Bantu, as we witnessed in this football competition, the Baka will gradually get to understand there is nothing to fear or be wary of with the Bantu. That feeling of familiarity and trust is very important and it can only be generated in common exercises like this football competition.

(A, Bantu player)

Intergroup cooperation is the third criterion set by Allport. Here, mutual dependence was reinforced by the idea that no one wins a football game by himself. To win games, Baka players needed to work together with Bantu players, and vice versa. Often, it was indicated that reconstructing this spirit of interdependence will be crucial to development in both Baka and Bantu communities in other aspects of life.

In order to develop, we need to work. In order to work, we need to be better organized and to be organized we need to show solidarity with one another. So, if we can continue to fellowship with the Bantu as we have done in the course of this football project, that will be the first step out of poverty for us.

(X, Baka player)

Union is strength and the solidarity between the Bantu and the Baka in this football programme is a synergy of this strength. Solidarity through football is a great approach to enable us get to know each other better and work together for the development of our villages.

(E, Bantu coach)

A last criterion mentioned by Allport is institutional support. This refers to consensus among the authorities and the relevant institutions about norms that support equality. In our action research, such support was palpable through the support from the Divisional Officer (administrative arm of the government of Cameroon) permitting the organization of our public event. Norms which support equality and impartiality were also reinforced in the selection of referees for the football games. In football, referees have the right to sanction players with yellow or red cards, stop or terminate the game due to risk factors and assess fouls and penalties. As such, it was agreed upon with the participants to select referees who did not belong to either of the local communities. As a non-Baka studying at a foreign university (and therefore with a high social status), the egalitarian standpoint was further strengthened by the main researcher treating both the Baka and the Bantu with the same respect.

A factor that was not mentioned by Allport, but which turned out to be decisive in our case study, is the place of contact. Both the choice of the venue where the football matches were played and the places where the latrines were dug were crucial factors in establishing solidarity between members of both communities.

The football matches were played on the primary school field in the heart of Lomie, a town predominantly inhabited by the Bantu. The latrines were dug with the help of Bantu men in the Baka villages. The choice of the Lomie town football field showed that no place is out of bounds for the Baka and that they are full members of the community. Likewise, the fact that Bantu were welcomed to help to dig latrines in the parts of the villages inhabited by Baka, demonstrated that their presence in these places is not met by fights, as many interviewees thought:

> I think the Baka do not have much tolerance when some occasion is organized next to their villages. If the football competition was organized on a field close to the Baka villages, there could have been fights.
>
> (BB, Bantu player)

> It felt really good to see the Baka recognized as part of the community in a public event like this in the heart of Lomie town.
>
> (Z, Baka player)

Conclusion

The key finding of this study is that, following displacement and the disruption of traditional forms of solidarity amongst the indigenous Baka, prolonged and profound encounters between Baka and Bantu in a mixed sports event can nurture forms of solidarity which transcend ethnicity and culture. By organizing a football programme coupled with the digging of communal latrines, the foundation was laid for prolonged contact between the Baka and the Bantu. After the intervention, representatives of both communities held more positive opinions about members of the other community. They were also more willing to help each other. In spite of the limited time frame, the achievements of this football programme will stay on the minds of the Baka and the Bantu in this area for a long time. The football shirts, trophies and medals they won, together with the latrines they were able to construct, will endure as constant reminders of the fact that Baka–Bantu relations were improved in the football programme and that intergroup solidarity had increased.

In doing so, the action research project we set up on the outskirts of the Dja Reserve also teaches us some lessons with regard to Allport's (1954) contact hypothesis. In fact, our action research project in East Cameroon approves many earlier studies claiming that sports activities that are not characterized by interethnic competition and that foster prolonged and profound engagement can challenge the stereotypes people of different ethnic backgrounds have about each other. While confirming many of the conditions originally proposed by Allport, from equal status and common goals to interdependence and institutional support, we also emphasized the importance of the place of contact in reducing prejudices between majority and minority group members. By focusing on practices of solidarity, we even made it clear that changes do not only occur in the attitudes and the discourses people hold, but also in the practices they deploy towards each other. In the villages where we conducted our study, such practices of solidarity are a counterweight for the erosion of intra-ethnic solidarities following the displacement of the Baka from the Dja Reserve and the establishment of unequal relationships of exchange that dominate the interactions between Baka and Bantu today.

Acknowledgements

We would like to express the deepest appreciation to the Baka and the Bantu of the Eastern periphery of the Dja Reserve who have shown positive attitudes by continually and unreservedly cooperating with regards to this research and scholarship. Our gratitude also goes to the dedicated local research assistants; Benjamin Kodju, Octave Ondoua, and Joseph Payong. We also appreciate logistics support from Helena Nsosungnine and GEOAID Cameroon (an international NGO for community development).

Notes

1 Because above all, digging of latrines in Baka villages was identified by the Baka as a community development initiative which will benefit the Baka more than the Bantu.
2 This decision to rule out Bantu participation by obligation was made by the Baka themselves.
3 The research team was composed of the first author, a Baka and a Bantu research assistant.
4 It is worth noting that no critical voices were raised in relation to the outcome of the football programme in improving relations between the Baka and the Bantu.
5 In the local lingo, the word 'different', or '*différent*' as in the French phrase *niveau différent*, is often used as a 'polite' way to imply hierarchy or superiority and inferiority.
6 Approximately 6USD.
7 Approximately 10USD.

References

Allport, G. W. (1954). *The nature of prejudice*. Reading, MA: Addison-Wesley.
Amin, A. (2002). Ethnicity and the multicultural city: Living with diversity. *Environment and Planning A, 34*(6), 959–980.
Awuh, H. E. (2016). *Conservation-induced displacement and resettlement: Building new bridges in social relations*. Unpublished PhD thesis, KU Leuven.
Bahuchet, S. (1985). *Les Pygmees Aka et la Foret Centrafricaine, Ethnologie ecologique* [The Aka pygmies and the Central Forest, ecological ethnology]. Paris: SELAF.
Bahuchet, S. (1990). Food sharing among the pygmies of Central Africa. *African Study Monographs, 11*(1), 27–53.
Bahuchet, S. (2000). *Les peuples des forêts tropicales aujourd'hui: 2. Une approche thématique* [The peoples of tropical forests today: 2. A thematic approach]. Brussels: APFT-ULB.
Barlow, F. K., Paolini, S., Pedersen, A., Hornsey, M. J., Radke, H. R., Harwood, J., Rubin, M., & Sibley, C. G. (2012). The contact caveat: Negative contact predicts increased prejudice more than positive contact predicts reduced prejudice. *Personality and Social Psychology Bulletin, 38*, 1629–1643.

Bayertz, K. (1999). Four uses of 'solidarity'. In K. Bayertz (Ed.), *Solidarity* (pp. 3–28). Dordrecht: Springer.

Brewer, M. B., & Miller, N. (1984). Beyond the contact hypothesis: Theoretical perspectives on desegregation. In N. Miller & M. B. Brewer (Eds.), *Groups in contact: The psychology of desegregation* (pp. 281–302). Orlando, FL: Academic Press.

Burger, J. M., Soroka, S., Gonzago, K., Murphy, E., & Somervell, E. (2001). The effect of fleeting attraction on compliance to requests. *Personality and Social Psychology Bulletin, 27*, 1578–1586.

Cernea, M., & Schmidt-Soltau, K. (2003). *Biodiversity conservation versus population resettlement: Risks to nature and risks to people*. Paper presented at the International Conference on Rural Livelihoods, Forests and Biodiversity, Bonn, Germany.

Cherry, F. (1995). *The stubborn particulars of social psychology: Essays on the research process*. London: Routledge.

Coffé, H., & Geys, B. (2007). Toward an empirical characterization of bridging and bonding social capital. *Non-profit and Voluntary Sector Quarterly, 36*(1), 121–139.

De Knop, P., & Elling, A., (2000). *Gelijkheid van kansen en sport* [Equal opportunities and sports]. Brussels: Koning Boudewijnstichting.

Dixon, J., Durrheim, K., & Tredoux, C. (2005). Beyond the optimal contact strategy: A reality check for the contact hypothesis. *American Psychologist, 60*(7), 697–711.

Dja Faunal Reserve. (2008). *The encyclopaedia of Earth*. United Nations Environment Programme-World Conservation, www.eoearth.org/view/article/151758/.

Dounias, E., & Froment, A. (2006). When forest based hunter-gatherers become sedentary: Consequences for diet and health. *Unasylva, 57*(2), 26–33.

Elling, A. (2004). 'We zijn vrienden in het veld'. Grenzen aan sociale binding en verbroedering door sport ['We are friends on the field'. Limits to social bonding and brotherhood through sport]. *Pedagogiek, 24*(4), 342–360.

Fischer, R., Callander, R., Reddish, P., & Bulbulia, J. (2013). How do rituals affect cooperation? *Human Nature, 24*(2), 115–125.

Hoffmann, A. (2002). Sozialintegrative Funktionen des Sports [Social-integrative functions of sport]. *Spectrum der Sportwissenschaften, 14*(2), 7–25.

Janssens, J. (1999). *Etnische tweedeling in de sport* [Ethnic divisions in sport]. Hertogenbosch/Arnhem: Diopter-Janssens and Van Bottenburg bv/NOC*NSF.

Jones, I. F., Young, S. G., & Claypool, H. M. (2011). Approaching the familiar: On the ability of mere exposure to direct approach and avoidance behavior. *Motivation and Emotion, 35*, 383–392.

Long, J., Hylton, K., Spracklen, K., Ratna, A., & Bailey, S. (2009). *Systematic review of the literature on black and minority ethnic communities in sport and physical recreation*. Leeds: Beckett University.

Matejskova, T., & H. Leitner (2011). Urban encounters with difference: The contact hypothesis and immigrant integration projects in eastern Berlin. *Social & Cultural Geography, 12*, 717–741.

Moreland, R. L., & Beach, S. R. (1992). Exposure effects in the classroom: The development of affinity among students. *Journal of Experimental Social Psychology, 28*, 255–276.

Ngoh, V. J. (1996). *History of Cameroon since 1800*. Limbé: Presbook.

Pettigrew, T. F. (1998). Intergroup contact theory. *Annual Review of Psychology, 49*(1), 65–85.

Pettigrew, T. F., & Tropp, L. R. (2005). Allport's intergroup contact hypothesis: Its history and influence. In J. F. Dovidio, P. Glick & L. A. Rudman (Eds.), *On the nature of prejudice. Fifty years after Allport* (pp. 262–277). Oxford: Blackwell Publishing.

Pettigrew, T. F., & Tropp, L. R. (2006). A meta-analytic test of intergroup contact theory. *Journal of Personality and Social Psychology, 90*, 751–783.

Pettigrew, T. F., Tropp, L. R., Wagner, U., & Christ, O. (2011). Recent advances in intergroup contact theory. *International Journal of Intercultural Relations, 35*, 271–280.

Podsakoff, P. M., MacKenzie, S. B., Lee, J. Y., & Podsakoff, N. P. (2003). Common method biases in behavioral research: A critical review of the literature and recommended remedies. *Journal of Applied Psychology, 88*(5), 879.

Ratna, A. (2010). 'Taking the power back!' The politics of British–Asian female football players. *Young, 18*(2), 117–132.

Schuermans, N. (2016). On the politics of vision and touch: Encountering fearful and fearsome bodies in Cape Town, South Africa. In J. Darling & H. F. Wilson (Eds.), *Encountering the city: Urban encounters from Accra to New York.* Farnham: Ashgate.

Shikongo, M. S. T. (2005). Report on threats to the practice and transmission of traditional knowledge regional report: AFRICA. Convention on Biodiversity. Retrieved from www.cbd.int/doc/meetings/tk/wg8j-05/information/wg8j-05-inf-03-en.pdf (accessed 24 November 2010).

Spracklen, K. (2013). *Leisure, sports & society.* Basingstoke: Palgrave Macmillan.

Stjernø, S. (2004). *Solidarity in Europe. The history of an idea.* Cambridge: Cambridge University Press.

Theeboom, M., Schaillée, H., & Nols, Z. (2012). Social capital development among ethnic minorities in mixed and separate sport clubs. *International Journal of Sport Policy and Politics, 4*(1), 1–21.

Valentine, G. (2008). Living with difference: Reflections on geographies of encounter. *Progress in Human Geography, 32*(3), 323–337.

Van der Meulen, R. (2007). Bier, Frikandel en Voetbal: lidmaatschap van sportverenigingen, vriendschappen, kenniskringen en veralgemeend vertrouwen [Beer, Frikandel and football: Membership in sports associations, friendships, knowledge networks and generalized trust]. *Tijdschrift voor sociologie, 28*(2), 166–191.

Verweel, P., Janssens, J., & Roques, C. (2005). Kleurrijke zuilen – Over de ontwikkeling van sociaal kapitaal door allochtonen in eigen en gemengde sportverenigingen [Colourful pillars – On the development of social capital by foreigners in their own and mixed sports clubs]. *Vrijetijdsstudies, 23*(4), 7–22.

Wessendorf, S. (2013). Commonplace diversity and the 'ethos of mixing': Perceptions of difference in a London neighbourhood. *Identities: Global Studies in Culture and Power, 20*(4), 407–422.

Wilshusen, P. R., Brechin, S. R., Fortwangler, C. L., & West, P. C. (2002). Reinventing a square wheel: Critique of a resurgent "protection paradigm" in international biodiversity conservation. *Society & Natural Resources, 15*(1), 17–40.

Zajonc, R. B. (1968). Attitudinal effects of mere exposure. *Journal of Personality and Social Psychology, 9*(2), 1–27.

9 Domesticating, festivalizing and contesting space

Spatial acts of citizenship in a superdiverse neighbourhood in Amsterdam[1]

Mandy de Wilde

Introduction

> It is a rainbow palette here, from well, Morocco, Cameroon, Brazil, Poland and many more countries. We don't have to travel the world as we all encounter it here in Slotermeer. The activities I visited this weekend are a good example of that. (. . .) On every street corner, every weekend, every weekday, there are so many more things here happening, all bursting with energy and inspiration. This weekend, there was a Surinamese Hindu event and Bosnian chess players played a game at the neighbourhood camp. And people even come from outside the neighbourhood to attend these activities, because it is so cosy here. (. . .) I feel this is the strength of what I see happening here. I find all those positive stories impressing and touching. All these small networks, all these interactions and meetings (. . .) lead to this colourful mosaic of people.

This vignette draws from a neighbourhood gathering organized as part of Neighbourhood Circle, a participation programme in a multicultural neighbourhood in Amsterdam, called Slotermeer. Sander,[2] a local administrator, is addressing residents. In his description of Slotermeer, he talks about there, namely 'Morocco, Cameroon, Brazil, Poland and many more countries' and how residents can encounter these diverse places and cultures in the here – 'every street corner' – and now – 'every weekend' and 'every weekday' – of the neighbourhood. More specific, he points to the fact that the here and now are enacted through actual encounters of residents – 'a neighbourhood camp', 'a Bosnian chess game' or 'a Surinamese Hindu event'. Thus, Sander introduces the neighbourhood as a place where diversity can be *encountered* through all kinds of activities. He also expects these encounters to provide a source for a cosy, multicultural atmosphere in the neighbourhood.

Sander's portrayal of Slotermeer's cosy ambiance is part of a local policy intervention which aims 'to enhance social cohesion' in the neighbourhood (Municipality of Amsterdam, 2009a). In order to reach that aim, a variety of encounters in public space are strategically provoked by Neighbourhood Circle community outreach workers who invite residents to initiate and organize informal, small-scale activities, such as weekly coffee mornings or dinners in a neighbourhood centre, annual street festivals or collective refurbishments of dilapidated playgrounds or squares. The policy assumption is that these activities provide for an encounter with diversity that will eventually bring about a fellow feeling and shared sense of public space among the diverse population of Slotermeer.

Neighbourhood Circle is not alone in expecting encounters to nurture social cohesion within the locality and public space of the neighbourhood. Many social scientists have argued that in 'superdiverse' cities (Vertovec, 2007) marked by a 'throwntogetherness' (Massey, 2005) or 'situated multiplicity' (Amin, 2008, p. 8) of cultures, lifestyles and ethnic backgrounds and where pre-existing shared norms and values coming from a long-shared history are absent as a source of social cohesion, such encounters provide a welcome means to bring and bind people together and reinforce a shared sense of public space. This type of civic formation has been qualified as 'conviviality' (Amin, 2008; Neal et al., 2013): a form of informal, pleasant interaction and cohabitation through everyday experiences and encounters with multiculturalism at the very local level. Often, conviviality is celebrated as an ideal form of civic formation in our present superdiverse cities:

> In recognition of the power of daily negotiation of difference in the workplace, public spaces, schools, housing estates and the like [. . .] conviviality is identified as an important everyday virtue of living with difference based on the direct experience of multiculture. [. . .] This interests stems from the recognition that the ethnography of encounter in the street and neighbourhood, school and workplace, park and square, is a crucial filter of social practice, affecting emotive, sensory, neurological and intellectual response towards both immediate others and the world at large.
>
> (Amin, A., 2008. Collective culture
> and urban public space. *City: Analysis of
> Urban Trends, Culture, Theory, Policy, Action*,
> *12*, 5–24, p. 18, reprinted by permission
> of Taylor & Francis Ltd, www.tandfonline.com)

However, Low (2006) criticizes the role conviviality plays in celebratory stories on everyday multiculturalism and shows that even despite pleasant interaction in public space, stereotyping, ethnocentrism and racism continue to take place. On another note, in her ethnography of a superdiverse London neighbourhood, Wessendorf (2014, p. 393) shows how conviviality is characterized by 'a fine balance between building positive relations across difference and keeping a distance' which leads to a 'civility with diversity' but not so much to a deeper engagement of residents with others in their surroundings. A critical approach to conviviality leads Amin to talk about conviviality as 'solidarity with space' (Amin, 2008, p. 18): a civic inculcation elicited by an understanding of urban public space as belonging in principle to everyone. He also points to the fact that conviviality can be felt and experienced in momentary encounters, but that it does not have to lead to a sustained mode of engagement with others:

The relatively unconstrained circulation of multiple bodies in a shared physical space [can be] generative of a social ethos with potentially strong civic connotations [but] the ethics of the situation [. . .] are neither uniform nor positive in every setting of throwntogetherness. The swirl of the crowd can all too frequently generate social pathologies of avoidance, self-preservation, intolerance and harm, especially when the space is under-girded by uneven power dynamics and exclusionary practices.

(Amin, 2008, pp. 7–8)

In this chapter I question if the encounters provoked by Neighbourhood Circle have a transcending potential to go from a mere solidarity *with space*, towards solidarity *with others*. I operationalize solidarity with others through the notion of 'acts of citizenship' (Işin, 2002, 2008). Acts of citizenship are deeds through which subjects fundamentally *relate* themselves to others and simultaneously *transform* themselves into citizens through 'ruptures in the given' (Işin, 2008, p. 25):

The moment of the enactment of citizenship, which instantiates constituents, also instantiates other subjects from whom the subject of a claim is differentiated. So an enactment inevitable creates a scene where there are selves and others defined in relation to each other.

(Işin, 2008, p. 18)

From this follows that acts of citizenship bring about a mode of engagement that differs from conviviality. When people position themselves as a citizen they develop an awareness of belonging and engagement through which they begin to reflect upon their identity in relation to others. So through acts of citizenship people act out *fundamental relationships* with others. Işin (2008, p. 19) argues that these relationships might enact a solidaristic mode of engagement as in being 'generous, magnanimous, beneficent, hospitable, accommodating, understanding, loving' towards others. However, these relationships might also enact agonistic modes of engagement, being 'competitive, resistant, combative, adverse' towards others or alienating modes of engagement, being 'vengeful, revengeful, malevolent, malicious, hostile, hateful' towards others (ibid., p. 19).

Below I describe what modes of engagement are born out of the encounters that are organized by residents in Slotermeer. I qualify an encounter as an act of citizenship if it brings about a rupture in the given. I argue that a rupture in the given on a neighbourhood level comes about when conventional dimensions between public and private become blurred. In her geography of the public realm, Lofland (1989) argues that the neighbourhood can be seen as 'parochial space'. This space is defined by 'a sense of commonality among acquaintances and neighbours who are involved in interpersonal networks that are located within "communities"' (Lofland, 1989, p. 455). For Lofland, parochial space link the intimate, private world of the household and kin networks to the public world of strangers and 'the street'. Yet, she also leaves room for permeating boundaries between this space, arguing that people's interactions and relationships bring this space into being and that changes in these relationships change the identity of the space itself:

> What is considered private, parochial or public space: whether a particular space is exclusive or inclusive; and whether that is, should be, may all be matters of conflict and/or negotiation.
>
> (Lofland, 1989, p. 470, 457)

Mere encounters have the potential to become acts of citizenship at the moment when residents make and use public space through deeds where private concerns and sensibilities are transformed into (temporary) public issues. Through an in-depth study of Neighbourhood Circle I focus on how residents negotiate the opportunity to use urban public space through the organization of encounters. I discern three groups of residents: immigrant women, the creative class and native Dutch residents and show how they, respectively, *domesticate* space, *festivalize* space and *contest* space. The results are encounters that ascertain agonistic, alienating and solidaristic modes of engagement between residents in a superdiverse neighbourhood.

Neighbourhood Circle: a participation programme in a superdiverse neighbourhood

Located on the outskirts of the city of Amsterdam, Slotermeer is a superdiverse neighbourhood (see Vertovec, 2007). A coalescence of diverse factors conditions the everyday life of residents: ethnicity, culture, education, class, origin and lifestyle. In 2010, with an increase of 16 per cent over the past decade, 59 per cent of its residents were officially categorized as 'non-Western migrants', of whom most are of Turkish, Moroccan or Surinamese descent, 9 per cent are categorized as 'Western migrants' and 32 per cent are listed as 'autochthonous', as being of Dutch descent (Municipality of Amsterdam, 2010).

The increasing superdiversity and the ongoing transitions in the neighbourhood are qualified in policy documents as an issue: it is said to cause 'an eroding sense of solidarity' among residents and to form a threat to the 'social cohesion' in the neighbourhood (Municipality of Amsterdam, 2009a). Initiated by the district government, Neighbourhood Circle was set up to target the above issues by engineering a community spirit in the neighbourhood. As face-to-face contact, friendly chats and intimate interactions were deemed the first steps towards kindling fellow feelings, a particular strategy was to transform public space in the neighbourhood into a cosy atmosphere and offer opportunities for encounters that could provoke this atmosphere (De Wilde, 2015).

I draw from data gathered from ethnographic fieldwork in the Neighbourhood Circle between 2009 and 2011. After a first stage of exploratory fieldwork I selected three projects for further consideration: 1) a weekly coffee morning organized by immigrant women in a neighbourhood centre; 2) an annual neighbourhood camp organized by the creative class; and 3) neighbourhood activities set up by native Dutch residents. First, immigrant women were surprisingly active within Neighbourhood Circle and made up one third of the volunteers on the list. Their activities were usually organized in the neighbourhood centre. Second, the creative class active within Neighbourhood Circle made up about 10 per cent of volunteers. They primarily organized outdoor events like a neighbourhood camp or urban gardening activities. Although small in number, their symbolic presence within Neighbourhood Circle was significant. They received a large share of the budget and their activities featured prominently in neighbourhood newspapers, in the Neighbourhood Circle newsletter, on the district website and in urban regeneration folders. Third, native Dutch volunteers had been qualified as 'the usual suspects' who organized activities 'spontaneously' and who, initially, took up the majority of the budget for Neighbourhood Circle (Municipality of Amsterdam, 2009b). In Neighbourhood Circle, native Dutch volunteers made up about a third of the volunteers, organizing dinners or coffee mornings for elderly at the neighbourhood centre, computer lessons, and often also engaged in the maintenance or re-development of playgrounds, squares, parks and other green space in the neighbourhood. These three groups represented the majority of residents who were active as volunteers within Neighbourhood Circle.

Over a period of two years I was a participant observer in a broad range of Neighbourhood Circle activities such as coffee mornings, dinners for the elderly and the neighbourhood camp. Furthermore, I attended Neighbourhood Circle gatherings and events, helped initiate and organize activities and conducted interviews with 16 policy practitioners, 18 immigrant women, 10 volunteers from the creative class and 12 native Dutch volunteers. I draw from field notes taken during these activities and events, from interviews, conversations and email correspondence with residents and policy practitioners as well as policy documents, weekly district newspapers, flyers, posters, websites and other communication material spread by the Neighbourhood Circle programme. In what follows, I describe how three groups of residents – immigrant women, the creative class and native Dutch residents – participate in Neighbourhood Circle. I conclude that Neighbourhood Circle does not so much provoke solidarities among the diverse population of Slotermeer, but rather among specific groups in which various others with different values and beliefs are included and excluded.

Domesticating space

The immigrant women I engaged with in Slotermeer are aged between 30 and 55; most are mothers with Moroccan, Turkish or Surinamese backgrounds. Some of them are first-generation immigrants, and left their country of birth many years ago. Others are second-generation immigrants born and raised in the Netherlands. They form a group of about 20 to 25 women, most of them unemployed. Due to unequal access to resources and opportunities (see Martin, 2002, on neighbourhood participation by women), they had not been publicly active in the neighbourhood before the advent of Neighbourhood Circle. While most had lived in Slotermeer for quite some time, they were not very familiar with (issues in) the neighbourhood. The knowledge, experience and use of the neighbourhood by these immigrant women was limited and can be understood as reflecting differences in access and mobility according to class, race and ethnicity as has been pointed out by human geographers (see Lynch, 1960, on mental mapping). Houda, a volunteer, explained that her mental map of the neighbourhood was prompted by her 'walking area': accessible by foot and bound by home, her children's school, grocery shops and the neighbourhood centre.

Many of the neighbourhood's immigrant women face social isolation. This is a social issue which Khadija and Houda, two volunteers who organize a weekly coffee morning for women in Slotermeer, have become aware of through the intervention of community outreach worker Wilma, who approached them during one of their visits to the neighbourhood centre and appealed to their sense of affinity and compassion with immigrant women like them who are suffering from social isolation. With Wilma's encouragement and support, Khadija and Houda organize a weekly opportunity for women in the neighbourhood to gather and share their personal stories in 'a safe environment'. Houda explains that she experiences the weekly coffee mornings 'like home, but without men and children'. However, what does Houda mean when she equates the neighbourhood centre with home?

Khadija's explanation offers a first point of reference. For her, the coffee mornings are 'a place for women of all cultures to come together':

> Here women can chit-chat, laugh, relax, forget and get information about everything that concerns women in the neighbourhood.

The importance attached to seemingly trivial activities such as laughing and relaxing illustrates how the coffee mornings allow women to engage in activities normally undertaken in the private realm of their homes or within the comfort of their intimate relationships. Furthermore, the reference to 'forgetting' points to the welcome ambivalence with respect to private problems related to family and household. Through laughing, relaxing and forgetting, these women engage in new intimate relationships with women of different ethnic and cultural backgrounds and express their feelings and concerns in the public environment of the neighbourhood centre. Women like Khadija and Houda, and the other women who organize and attend the coffee mornings, *domesticate* the neighbourhood centre: through their home-making practices, the public space of the neighbourhood centre acquires comforting, relaxing qualities because it becomes a sort of living room, where some women can share emotions and experiences that are closely related to things happening in the privacy of their households or families.

One week, something unexpected occurred during the coffee morning which illustrates the ambiguity of this domestication of public space. Community outreach worker Wilma had sent out invitations for the coffee morning to the neighbourhood's elderly, but had forgotten to mention that only women are allowed. At one point, an old man enters the room with an invitation in hand. Suddenly, panic stirs in the room, as two women rush to put on their headscarves, and others nervously straighten their headscarves around their faces. Wilma, of native Dutch descent, welcomes the old man, while Aisha angrily utters words in her native language; Emine does not understand the tumult and asks Aisha to talk in Dutch, which Aisha, upset, fails to do. Here, the private sphere that is acted out confronts the public function of the neighbourhood centre. Community outreach worker Wilma tells the women that a neighbourhood centre is a public space and is 'for every-body'. Emine emphasizes this as well as she tries to engage Aisha in conversation. Yet, Aisha and two other women leave; some of those who stay remain uncomfort-able. Some start doing the dishes, while others start talking to the old man who appears flabbergasted by the stir he has caused.

For some women, a transgression occurs that is experienced as out-of-place and they fundamentally relate themselves to this other person. They do not recog-nize the man as a neighbour but as a stranger and a threat to their gendered sphere of belonging. Unexpectedly, the entrance of the man becomes a reminder of the diversity among the women, as his presence affects them differently. As such, this 'strange encounter' (Ahmed, 2000) establishes adverse reactions and an agonistic mode of engagement between some of the women and the man. Yet, there are also women who accommodate the man, start to engage in conversation and act out a solidaristic engagement with him as they are willing to share the public space of the neighbourhood centre with him.

Moreover, this rupture in the given revives a discussion on what immigrant women want and they engage collectively in a reflection of their group iden-tity. Out of the discussions which follow over the course of the following weeks, it becomes clear that they see the coffee morning as an extension of their own homely living room and, therefore, it is important to gather without men be(com)ing present. Women like Emine who do not mind the attendance of men accom-modate the feelings and wishes of the women who do mind, and together, as a group, they decide it is time to get a place of their own and apply for funding to realize this women's centre. As such, through this application, a solidaristic mode of engagement is enacted among these women, because women reflect upon their identity in relation to others and what they want to share with others. Yet, it also enacts an agonistic relationship with male others because the women simulta-neously aim to separate themselves from their male neighbours as their desired centre is not open to everybody. Doing so, they make a territorial claim to neigh-bourhood space.

Festivalizing space

Apart from immigrant women, other residents also organize encounters in public space. Every year in June, residents of Slotermeer have the chance to go on a 'mini-holiday' in their own living environment. This festival – called 'neigh-bourhood camp' – consists of a two-day programme that invites residents to experience diversity in all its forms. The festival features a wide array of activities organized around music, food, performance and dance, among others an 'Oriental belly dance, Tango guitar music, a henna tattoo workshop, a traditional Greek dance performance, a traditional Chinese lion dance, traditional Javanese court dance, Japanese calligraphy, yoga, Old Dutch games and a Turkish food work-shop'. An emphasis on folklore runs through the whole programme: it shares and celebrates the creative expressions and traditions of everyday life in different ethnicities and cultures.

Chris, a highly educated professional who works in the creative sector of Amsterdam, is the initiator of neighbourhood camp. He felt 'a bit lost and uncom-fortable, not really at home' when he came to live in Slotermeer over a decade ago. Not knowing most of his neighbours and noticing, to his discomfort, that the streets were covered in litter he decided to change that. He organized a neighbour-hood event which over the years became known as the 'neighbourhood camp'. By 2010 Chris had gathered a core group of about 20 volunteers who can be described as 'pioneer gentrifiers' (Butler & Lees, 2006): with two exceptions, the volunteers all belonged to the so-called 'creative class' (Florida, 2002). They were highly educated professionals of Dutch descent working in the arts, design, media and knowledge-based sectors and most of them came into the neighbourhood before the urban regeneration of Slotermeer was well underway. What distinguishes them from the group of native Dutch residents in Slotermeer is that they do not so much identify as native Dutch or authochtonous, but rather as cosmopolitans, as free from national ideas, prejudices and attachments.

The fact that Slotermeer carried a public stigma of a disadvantaged neighbour-hood had not stopped these volunteers from moving in. Quite the contrary; to Harold, an artist, the diversity of ethnicities, cultures and lifestyles in Slotermeer added to his feeling of living in a 'laboratory'. The camp offered Harold an opportunity to 'invest' his 'creative talents' for the benefit of 'reaching out and connecting' with other residents. To him 'doing things together strengthens the bonds in this multicultural neighbourhood' and in making this statement he mim-ics the aim of Neighbourhood Circle. Carin, a journalist and designer, resonates the importance of cosiness as she explains her motivation to become involved in neighbourhood camp:

> I think it is good to do something nice for your neighbourhood. You know, Geert Wilders [a Dutch right-wing politician] can say that festivities like these are rubbish. Well, just come and see. The neighbourhood camp is very cosy. Fatima and Henk and Ingrid[3] might not go on an actual holiday together, but that's not what it's about. It's about getting to know each other informally and in a cosy way. About creating a nice atmosphere in the neighbourhood (. . .). If I am new to a place I get homesick quite easily, I will feel a bit *unhe-imlich*. I have to feel good when I walk the streets here.

Harold and Carin describe the ethnic and cultural diversity in Slotermeer as exciting and a means for inspiration, but simultaneously also feel they have to appropriate it in order to feel less *unheimlich* and more familiar in their new living environment.

The neighbourhood camp tries to help residents get acquainted with their neigh-bourhood differently by transforming a small, green park into a colourful, vibrant and musical campsite. Through an abundant stimulation of the senses – the setting,

music, food and performances – the camp offers opportunities for temporary interactions where those who go on holiday can behave spontaneously, where they
can leave their worries behind and where they can enjoy a 'cosy' and 'respectful atmosphere' – demarcated by a colourful 'customs post' which symbolically
marks the entrance to the field.

In doing so, volunteers from the creative class *festivalize* space: they imbue a
green park with aesthetic attributes, cheerful activities and conspicuous performances and re-imagine it into a temporary conviviality. They transform urban
public space into a place that resembles the everyday life world of those volunteers from the creative class who organize the camp. It is a world characterized
by art and culture and not so much dictated by social issues that characterize
the everyday reality of a majority of residents in this superdiverse and deprived
neighbourhood, like bad housing conditions, the upcoming demolition of housing blocks, poverty or social isolation. Hence, these volunteers actively try to
make social problems and everyday struggles temporarily absent. This is a conscious decision. Chris explains that the volunteers abandoned the original idea
of cleaning the street:

> We thought that the aspect of art, culture, eating and drinking was much nicer
> than the aspect of cleaning the streets. It just expressed certain negativity, like
> that the government isn't doing its job properly. Now it is much more about
> celebrating and coming together. It is also what we like to do and enjoy.

Neighbourhood camp thus offers residents of Slotermeer an opportunity for informal, pleasant interactions through an encounter with culture in its most diverse
forms. Doing so, these volunteers act out a cosy, multicultural conviviality just
like the Neighbourhood Circle programme intends. It is a conviviality which
offers Chris, Carin and Harold a stage to express and share their talents, interests
and dreams for Slotermeer publicly.

Despite the fact that Chris and his fellow volunteers have decided to focus only
on the positive aspects of living in Slotermeer, they are very aware of some of the
issues the neighbourhood faces. Chris expresses his discontent about residents
who are 'too indifferent to look after their living environment and don't commit
to the neighbourhood'. He talks about 'the people across the street', who live in
'the social housing over there', who have set up a marijuana plantation and get
into loud fights during the night. He also thinks the large influx of temporary residents is problematic. He relates his discontent to the upcoming demolition plans,
saying 'at a certain point it makes sense that they tear down those houses'. 'Those
houses' are small, social housing blocks that are poorly maintained opposite
Chris's block, which consists of large, owner-occupied dwellings with a garden.
But the most pressing issue for Chris appears not to be the presence of these residents and the bad housing conditions they live in, but the *absence* of something
else, which relates to their presence. He does not perceive the demolition plans as
a threat; for him it provides an opportunity for a 'cultural impulse to the neighbourhood'. He relates the current population of Slotermeer to the poor cultural

infrastructure in the neighbourhood and believes that the influx of new people will actually benefit 'people like me' as it will most probably lead to a diversification of shops, restaurants and cafes in the neighbourhood:

> At the moment, there are no interesting cafes here; you will have to go downtown for that. I do go to the Pink New West drinks [monthly happenings for gay and queer people in the district], but it is not a nice cafe or anything like that. (. . .) And all those greengrocers and similar looking Blokker-like [Dutch discount store] stores with the same plastic crap. There is just no policy about that now. There isn't a single bookshop in the neighbourhood. Not one! No quality shops, no cultural centres. There is one artists' club in [an adjacent neighbourhood] with a restaurant. It's a bit alternative, a bit posh. It's really nice. There should be something like that in the heart of our neighbourhood. Or an art-house cinema, like what's happening now on the Westergasterrein [a popular cultural space in Amsterdam]. Because we are in the process here of drawing more groups to the neighbourhood. Slotermeer has the image of a poor people's district, but people with money and taste also live here. There isn't enough for them here.

Through his reference to people with 'taste', he indicates the future residents he is expecting: highly educated, with a lot of cultural capital, just like him and his fellow volunteers. Interestingly, Chris uses the personal pronoun 'we' when talking about the ongoing urban regeneration and transformation of the neighbourhood which is designed and developed by the municipality of Amsterdam, the district administration and housing associations. He perceives the camp as a partner in this transformation. Once, during a meeting, a volunteer who cannot be qualified as a member of the creative class, asks whether the camp should not show the 'Sadness of Slotermeer', referring to a documentary made by a local TV station that depicts the severe consequences of demolition plans in the neighbourhood on the daily life of some residents. The volunteer is swiftly silenced by Chris who states that 'we will not talk about politics'. The other volunteers who are present agree. These quotes illustrate how politics, as perceived by Chris and most of his fellow volunteers, are actively kept out of the neighbourhood camp. However, this act of avoiding voice can be perceived as a political act in itself because it helps these volunteers to enact, present and promote a gentrified image of Slotermeer that is in line with the governmental plans for a future Slotermeer.

So the festivalization of the park undertaken by these volunteers results in encounters that do not give rise to *fundamental* ways of relating or a sustained mode of engagement between neighbours: these volunteers just want to have a good time and are perfectly comfortable with communicating with various types of people, but many of them know little about the way of life of others or attempt to know more. Neighbourhood camp enacts a 'cool conviviality' or 'light engagement' (Neal et al., 2013, p. 318) that neither gives life to alienating, agonistic or solidaristic modes of engagement between the volunteers and their neighbours.

Contesting space

Whereas immigrant women and the creative class relatively recently started to organize encounters in public space, Slotermeer is also populated by a group of native Dutch residents, mostly aged 55-plus and mostly from a working-class background, who have been active in public space for over a decade. To them, new encounters in public space function as a podium showcasing the socio-demographic and socio-cultural transformations that have taken place not only in the neighbourhood, but in the city and the country as a whole. These native Dutch residents perceive themselves as a minority in their own neighbourhood and feel uprooted. In an interview, Frans, chairman of a neighbourhood committee, voices their concerns:

> The interethnic relations are still an issue in this neighbourhood. There are older Dutch people that have the feeling of being 'lost in familiar places'. They ask themselves: 'Am I walking in my own country or through the Kasbah of Tangier?'

These native Dutch volunteers display a strong social attachment to other native Dutch in the neighbourhood. In an interview Dirk, a retired widower, explains why he participates in Neighbourhood Circle. He shares his stories about growing up in a small village and how he initially missed the 'solidarity of a small village' when he moved to Slotermeer:

> As a volunteer I was also looking for that same feeling. It brings you into contact with people who think alike or who might fight against the same things. Or fight for something alike of course. It creates a bond.

In Slotermeer, I engaged with a group of 20 native Dutch volunteers, the majority male. These volunteers attempt to *contest* public space through the organization of encounters in the neighbourhood.

To native Dutch volunteers, the multicultural composition of Slotermeer is perceived as an issue. When Frans refers to himself and his native Dutch neighbours as tourists in their own country, his words resonate Dutch national public debate in which politicians from both right- and left-wing parties refer to the native Dutch in disadvantaged neighbourhoods as 'foreigners' who feel lost in their own living environment (Duyvendak, 2011, pp. 84–105):

> *Frans:* Last Sunday, the minister of integration made clear in [a TV programme] that there is a lot of hurt and feelings of abandonment among autochthonous Dutch people who, for decades, have had the feeling that they didn't matter anymore. I mean, a good acquaintance of mine had it happen to her. She goes to a club in the neighbourhood centre. She is white, but if she goes and gets coffee, she has to pay for it, but a Turkish or Moroccan woman who's also standing in line doesn't have to pay.

MdW: I don't understand. . .

Frans: Well, I don't understand it either. But she thinks, dammit, why do I have to pay and she doesn't? Yes, because she's an immigrant. But yes, it happens. And those are the things that keep stirring ill-feelings and keeps on feeding a fire between ethnic groups.

Frans displays how his sensitivity to ethnic and cultural others is formed, on the one hand, by public and political discourses and, on the other hand, by reflections and gossip about everyday encounters in Slotermeer which travel by word of mouth. The imagined Other – the immigrant woman – is constructed from watching television: it is an Other who is Muslim, and more importantly, not yet integrated well enough into Dutch society and culture.

When asked about the tense interethnic relations in the neighbourhood, Dirk refers to the district administration in order to explain his own feelings of discontent about the troubled social relations:

> Do you know who have made us aware of those differences and made those dividing lines? The district administration. (. . .) Whenever public encounters or events were organized by them, there are always Moroccan snacks and always the same group of Moroccans providing the food. (. . .) The Moroccans appeared to be close to the district administration and always got their way, while we had to struggle to keep our activities going.

Dirk and his fellow volunteers perceive the diversity and the accompanying public recognition of multicultural encounters as problematic. To them it affects their ability to publicly articulate their Dutch identity through practices such as organizing encounters and receiving public recognition for it.

The use of squares for activities that are qualified as unfamiliar, or not contributing to Slotermeer, is also contested by native Dutch volunteers. Rob, a volunteer who organizes coffee mornings and activities for elderly people, remarks on what he describes as a 'multicultural festival', which took place on a public square in the neighbourhood:

> There are these Moroccan residents who organize festivals for the neighbourhood. They say it's for the neighbourhood, but there are only Moroccan acts. They flew in a belly-dancer from Morocco, who is dancing for 15 Moroccans at the cost of 5,000 euro. Well, it's an asset for the neighbourhood, really, it really brings people together [his tone is sarcastic]. It has been like that before, we complained about it [to local administrators and community outreach workers]. I mean, mind you they colonize the whole square. And, you know, Plein 40–45 is a big square. I heard that nowadays they adjust the programme and hire an accordion player as an in-between act to please the elderly residents. What do they think? We will attend somewhere in-between the rap workshop and the Koran workshop? I don't think so.

The word 'colonize' is used more often by people like Rob. Together with addressing some of the other residents as 'Moroccans' or 'foreigners', this indicates that, in the eyes of Rob and his fellow native Dutch, these residents do not really belong to Slotermeer. Volunteers who celebrate their version of a multicultural conviviality on a public square – and are able to publicly and collectively manifest themselves – transform a familiar public place into something unfamiliar. This state of affairs affects the urban public space as experienced by people like Rob. The perceived presence of others plays an important role in giving meaning to the use and experience of public space and to them, forms an important indication of the quality of public space in the neighbourhood.

Subsequently, this perception gives rise to new, and strengthens existing, alienating modes of engagement with ethnic and cultural Others. They actively oppose the idea of a cosy, multicultural conviviality, desired by the Neighbourhood Circle programme, through *contesting* space: through their encounters they articulate something that they experience as being in public decline in the neighbourhood, namely Dutchness. They use elements from popular Dutch culture to create a space in which they can publicly articulate their Dutchness. Dinners are organized with *stamppot maaltijden* (mashed potatoes), afternoons with old Dutch games and traditional cultural craft workshops, like making clogs. There are also 'Amsterdam' afternoons and evenings. The adjective 'Amsterdam' refers to the music played, 'tearjerkers', a subgenre with simple melodies and lyrics in Dutch, generally about love, family relationships and mundane life. Though it is not voiced as a conscious strategy, organizing 'Amsterdam' afternoons in the neighbourhood centre is a way of indulging in togetherness through excluding others, in this case immigrants. This is also recognized by Jasper, a community outreach worker:

> Well, of course Mohammed from around the block doesn't attend an Amsterdam afternoon. Not every voluntary activity means something to the whole community.

These volunteers also indulge in togetherness through the use of food. Some native Dutch volunteers make clear that their dinners are intended only for a certain kind of people. Again, they oppose themselves to an Other they do not want to come close(r) to. On paper, an invitation, Christmas dinners or 'mashed potato dinners' are open to everybody. This is the first thing Rob, a volunteer who organizes such dinners for the elderly, tells me when we are drinking coffee in a small neighbourhood living room run and supervised by him. I ask him who attends his dinners. Initially, he states that they are open to 'lonely elderly', but at a later point in our conversation he refers to the issue of religion stating that, of course, the food is not going to be 'halal'. Also, he remarks that 'you cannot take away a tasty lard cutlet from our elderly' as, to him, a lard cutlet is part of the traditional Dutch mashed potato dish. His 'our' refers to those elderly people of Slotermeer who, according to him, 'built up our society after the war and that deserve a couple of cosy afternoons and evenings for their services for our country'. By putting an emphasis on lard cutlet – a dish prepared with pork – Rob indicates that he is not going to adapt to Islamic customs, as many Turkish and Moroccan residents do not eat pork for religious or cultural reasons.

To summarize, during dinners and other encounters organized by autochthonous volunteers, the music and food – no halal food, the use of pork and the playing of Dutch tearjerker music – plays a big role, articulating a sensual Dutch presence in public space. So, while Neighbourhood Circle invites them into an informal, pleasant interaction with other residents, native Dutch volunteers do so, but through fundamentally relating to those they perceive as Other. They enact alienating modes of engagement with their ethnically and culturally different neighbours, which are part of a strong solidaristic mode of engagement with their fellow native Dutch neighbours.

Conclusion: solidarities and spatial acts of citizenship

Neighbourhood Circle, a participation programme in an Amsterdam neighbourhood, offers its heterogeneous population the opportunity to organize encounters with the aim of bringing and binding people together and reinforce a sense of shared public space. An ethnographic insight in these encounters shows how residents negotiate this opportunity for a 'civic becoming' (Amin, 2008, p. 22) or 'conviviality' (Amin, 2008; Neal et al., 2013). At times, some residents engage in 'acts of citizenship' (Işin, 2008) where a *fundamental* relation to their neighbours is enacted – a political becoming. This is galvanized in and through encounters where private and personal concerns and sensibilities are transformed into public issues.

First, immigrant women *domesticate* space: in a weekly coffee morning, immigrant women relocate the domestic ethics of private practices and personal concerns which are part of the private sphere of the household or the intimate sphere of the family to a neighbourhood centre. The neighbourhood centre is temporarily privatized so that immigrant women can feel comfortable and at ease there. Doing so, these women cultivate existing, and forge new, bonds of affinity with women from different ethnic and cultural backgrounds in the neighbourhood. However, this gendered solidarity appears back-to-back with feelings of otherness. When, accidentally, a man enters the coffee morning, the agonistic aspects of this temporary, gendered solidarity become tangible. The accidental encounter stirs up a discussion. Subsequently, the women discuss what binds them together and reflect upon what they need in opposition to what the participation programme needs from them as active citizens. In order for their gendered solidarity to flourish, they need a space in which particular others (men) are not allowed. The example of requesting their own space is an act of citizenship which arises out of a confrontation with otherness and it demonstrates a tension between group formation and the publicness of urban space. Paradoxically, this gendered solidarity and act of citizenship would diminish the public character of neighbourhood space. Once again, the personal is made political, albeit in a different manner than feminists intended when first using the slogan (Lister, 1997).

Second, volunteers from the creative class *festivalize* space: through an annual neighbourhood camp they imbue a small public park with aesthetic attributes, creative activities and conspicuous performances and re-imagine this public space into a multicultural spectacle that resembles their personal preferences and concerns (see Johansson & Kociatkiewicz, 2011). In these frivolous encounters residents from diverse walks of life mingle: immigrant parents and their children share a table or activity with young urban professionals. The neighbourhood camp shows that these pioneer gentrifiers do not fundamentally relate to their neighbours either. Rather, these volunteers engage in a 'civility towards diversity' (Wessendorf, 2014, p. 394): a fine balance between building positive relations and keeping a distance. The bulk of the activities and interactions focus on commonalities rather than differences and avoid addressing possible tensions or issues in the neighbourhood. Neal et al. (2013, p. 318) describe these 'mundane competencies for living cultural differences' as 'light engagement'. These light solidarities resemble what Amin (2008, pp. 18–19) calls 'conviviality': conviviality is identified as an important everyday virtue of living with difference based on the direct experience of multiculturalism, getting around the mainstream instinct to deny minorities the right to be different or to require sameness or conformity from them. This conviviality is characterized by both avoidance of deeper contact and engagement with others.

Finally, native Dutch residents *contest* the use of public space by immigrants: they experience it as an unfamiliar use of neighbourhood centres and parks and it signals the erosion of their traditional way of life. The living environments of these native Dutch residents have been directly affected by globalization and immigration. Having moved into the neighbourhood decades ago, many of them are especially aware of the change in the ethnic and socio-economic composition of their neighbourhood which, in their perception, co-occurred with their social and cultural marginalization and an increasing 'disorder' in public space (Savage et al., 2005; Sampson, 2009). As such, their encounters are an attempt to preserve their place attachment to the neighbourhood and make present what they feel they are losing, namely a personal attachment that they can publicly articulate, express and live out through activities in public space and community centres. Not only do they 'feel lost in familiar places', but these public places have actually become unfamiliar as they look, sound, smell and are used differently by newly arrived residents into the public space: immigrant women and the creative class. The light and gendered solidarities of these residents exclude native Dutch residents who are not able or willing to identify with these relationships, as to them, it feels unfamiliar. In response, they publicly articulate Dutchness and this exclusive identity is intertwined with their perception of the local identity of Slotermeer. As such, they present themselves as natives and weave issues related to nationality and locality into an articulation of Dutchness (see Van Reekum, 2012) and enact solidaristic relationships among their fellow native Dutch residents. Through this public articulation of Dutchness, political meanings about ethnic groups become part of the social relations in Slotermeer as is demonstrated in their alienating engagement with immigrant women; they refuse to share public space with them.

To conclude, Neighbourhood Circle provokes not so much solidarity among the diverse population of Slotermeer, but solidarities among specific groups in which various others with different values and beliefs are included and excluded. These relationships emerge through encounters which are all about the transformation of space. Sometimes, these encounters bring about a mode of engagement which goes beyond mere conviviality. It is through these 'spatial acts of citizenship' that immigrant women and native Dutch volunteers develop an awareness of belonging and engagement through which they begin to reflect upon their identity in relation to others and act out *fundamental relationships* with others. In Slotermeer, these spatial acts of citizenship are generative of a public space that is 'liminal' (Buckingham et al., 2006): the everyday, the intimate and the personal become relational places where the agency of immigrant women, the creative class and native Dutch residents connects the public and the private. How much space different groups should enjoy to enact solidarities – on streets, parks and squares or in neighbourhood centres – remains a burning issue. Active liminality in order for everyone to use public space remains a delicate balancing act, as manifestations of one group's familiarity, feelings of home and engagement may be met by antagonism and alienation from other groups.

Notes

1 This chapter contains revised extracts from the following article and book chapter:

> De Wilde, M. (2016). Home is where the habit of the heart is. Governing a gendered sphere of belonging. *Home Cultures: The Journal of Architecture, Design and Domestic Space*. DOI: 10.1080/17406315.2016.1190583.
> De Wilde, M. (2015). Profound coziness. Affective citizenship and the failure to enact community in a Dutch urban neighbourhood. In *Die Ambivalenz der Gefühle. Uber die verbindende und wiederspruchliche Sozialitat von Emotionen* (pp. 125–143). Wiesbaden: Springer Fachmedien.

2 Fictitious names guarantee the anonymity of my informants.
3 Dutch right-wing politician Geert Wilders uses the names 'Henk and Ingrid' in national public and political debate, as symbolic of 'typical Dutch' people who do not feel at home in a multicultural society. As such, it has become a well-known expression in Dutch public debate, media and everyday speech.

References

Ahmed, S. (2000). *Strange encounters: Embodied others in post-coloniality*. Hove: Psychology Press.
Amin, A. (2008). Collective culture and urban public space. *City: Analysis of Urban Trends, Culture, Theory, Policy, Action, 12*, 5–24.
Buckingham, S., Marandet, E., Smith, F., Wainwright, E., & Diosi, M. (2006). The liminality of training spaces: Places of private/public transitions. *Geoforum, 37*, 895–905.
Butler, T., & Lees, L. (2006). Super-gentrification in Barnsbury, London: Globalization and gentrifying global elites at the neighbourhood level. *Transactions of the Institute of British Geographers, 31*, 467–487.
De Wilde, M. (2015). *Brave new neighbourhood. Affective citizenship in Dutch territorial governance*. Unpublished doctoral dissertation, University of Amsterdam.

Duyvendak, J. W. (2011). *The politics of home: Belonging and nostalgia in Western Europe and the United States*. Basingstoke: Palgrave Macmillan.

Florida, R. L. (2002). *The rise of the creative class. And how it's transforming work, leisure, community and everyday life*. New York: Basic Books.

Işin, E. (2002). Engaging, being, political. *Political Geography, 24*, 373–387.

Işin, E. (2008). Theorizing acts of citizenship. In E. Işin & G. M. Nielsen (Eds.), *Acts of citizenship*. London: Zed Books.

Johansson, M., & Kociatkiewicz, J. (2011). City festivals: Creativity and control in staged urban experiences. *European Urban and Regional Studies, 18*, 392–405.

Lister, R. (1997). *Citizenship: Feminist perspectives*. New York: New York University Press.

Lofland, L. H. (1989). Social life in the public realm. A review. *Journal of Contemporary Ethnography, 17*, 453–482.

Low, S. (2006). The erosion of public space and the public realm: Paranoia, surveillance and privatization in New York City. *City and Society, 18*, 43–49.

Lynch, K. (1960). *The image of the city*. Cambridge, MA: The Technology Press and Harvard University Press.

Martin, D. G. (2002). Constructing the 'neighbourhood sphere': Gender and community organizing. *Gender, Place and Culture, 9*, 333–350.

Massey, D. (2005). *For space*. London: SAGE.

Neal, S., Bennet, K., & Cochrane, A. (2013). Living multiculture: Understanding the new spatial and social relations of ethnicity and multiculture in England. *Environment and Planning C, 31*, 308–323.

Sampson, R. J. (2009). Disparity and diversity in the contemporary city: Social (dis)order revisited. *British Journal of Sociology, 60*, 1–31.

Savage, M., Bagnall, G., & Longhurst, B. (2005). *Globalization and belonging*. London: SAGE.

Valentine, G. (2008). Living with difference: Reflections on geographies of encounter. *Progress in Human Geography, 32*, 323–337.

Van Reekum, R. (2012). As nation, people and public collide: Enacting Dutchness in public discourse. *Nations and Nationalism, 18*, 583–602.

Vertovec, S. (2007). Super-diversity and its implications. *Ethnic and Racial Studies, 30*, 1024–1054.

Wessendorf, S. (2014). 'Being open, but sometimes closed'. Conviviality in a super-diverse London neighbourhood. *European Journal of Cultural Studies, 17*, 392–405.

List of policy documents

Municipality of Amsterdam. (2009a). City district Geuzenveld-Slotermeer. *Ideeën zonder eigenaar. Hoe [Neighbourhood Circle] vatbare bewoners gaat aanjagen om hun schouders ook echt onder ideeën te zetten [Ideas without owners. How Neighbourhood Circle will encourage residents to engage with ideas]*.

Municipality of Amsterdam. (2009b). City district Geuzenveld-Slotermeer. *Jaarrapportage: Budget Bewonersinitiatieven Wijkaanpak [Annual Report: Budget Citizeninitiatives Neighbourhood Renewal Policy]*.

Municipality of Amsterdam. (2010). Research and Statistics Department. *Staat van de wijk III. Geuzenveld-Slotermeer [Condition of the Neighbourhood III. Geuzenveld Slotermeer]*. Retrieved from website of the Research and Statistics Department: www.os.amsterdam.nl/nieuws/download/711/2010_svdw_geuzenveld_slotermeer.pdf

10 Afterword

Solidarities, conjunctures, encounters

David Featherstone

Introduction

The fine Scots ballad *Erin Go Bragh* recounts the story of an encounter between a Highlander Duncan Campbell, 'from the shire of Argyll', and a policeman on the streets of Edinburgh.[1] The terms of the encounter are set early on by Campbell, the song's narrator. He recalls how the policeman 'glowered in my face and he give me some jaw, saying when come ye over from Erin Go Bragh?' [Ireland]. Pivoting on the misrecognition of Campbell as an Irish immigrant, 'I knew you a Pat [Irish] by the cut of your hair/ but you all turn to Scotsmen as soon as you're here', the song challenges two forms of prejudice which have structured Scottish society; that against Highlanders, and against the Irish, hostility which has often been articulated through anti-Catholicism.

What makes the song significant is how this misrecognition is negotiated. Campbell refuses to join in the demonization of the Irish, 'Well were I a Pat and you knew it were true/ Christ were I the devil well what's that to you', instead turning his ire on the prejudiced policeman. 'The Polis' in turn meets his come-uppance in the form of spirited resistance: 'a lump of black thorn that I held in my fist/ across his big body I made it to twist/ and I showed him a game played in Erin Go Bragh'. The song ends with a rousing declaration against prejudice: 'So come-all- ye young people- wherever you're from, and don't give a damn to which place you belong'.[2]

A nineteenth-century song,[3] though one still common in the repertoire of Scottish and Irish singers, might seem an arcane reference point for contemporary debates about the relations between multi-ethnic politics, place and solidarity. The song, however, powerfully speaks to three key issues which will be central to this afterword, and which are integral to thinking about multi-ethnic politics in emergent conjuncture(s). First, it locates encounters in relation to uneven geographies, bringing the terms on which they become articulated in particular places into contestation. Second, while recent literatures have often staged encounters as rather anodyne occurrences, the song emphasizes that they can be freighted with significant histories and geographies of contestation. They can, as it makes clear, become politicized and rendered antagonistically. Finally, the song's refusal of chauvinistic identity and exclusionary notions of place demonstrates that articulations of solidarity can be produced through such struggle.

Through its rigorous analysis of what is at stake in thinking about the relations between place and the construction of diverse solidarities, this volume makes a welcome contribution. As the chapters deftly illustrate, there are, fortunately, diverse modalities of encounter and solidarity through which uneven relations can be negotiated, and there usually exist alternative strategies to recourse to than the rather drastic means of hitting a cop with a 'blackthorn stick'. This afterword develops these concerns with the political construction of place by locating such debates in dialogue with the uneven relations between past and present. I then draw out the important contribution the chapters in the book make to thinking about the diverse modalities of solidarity and encounter and argue that this is a significant and timely project. I conclude with some observations about the pressures on, and necessity of, constructing diverse solidarities in the current conjuncture marked by crisis and austerity.

Historicizing encounters and the political construction of place

Stuart Hall (2000, p. 218) has written of the post-Second World War 'migration into Britain from the non-white global periphery' that 'the pathways which these migrants followed' were 'marked out' by the 'relations of colonization, slavery and colonial rule' that linked 'Britain with the empire for over 400 years'. He contends:

> These historic relations of dependency and subordination were *reconfig-ured* – in the now classic post-colonial way – when reconvened on domestic British soil. In the wake of decolonization and masked by a collective amnesia about, and systematic disavowal of 'empire' (which descended like a Cloud of Unknowing in the 1960s), this encounter was interpreted as 'a new beginning'. Most British people looked at these 'children of empire' as if they could not imagine where 'they' had come from, why, or what possible connection they could possibly have with Britain.
>
> (Hall, 2000, p. 218, emphasis in original)

This process of reconfiguring relations, Hall argues, involved challenges to the 'settled notion of British identity' and posing 'the multi-cultural question'. In this way, he stresses the importance of thinking about the histories and geographies, the 'conjunctures', through which encounters are experienced and constituted. Such geographies were directly imbricated with Britain's 'long, disastrous, imperialist march across the earth' (Hall, 1988, p. 172). By doing so, Hall emphasizes that the 'multi-cultural' question is not a dispassionate, technical question about living with difference. Rather, he locates multi-culturalism in relation to antagonistic ways of envisioning societies and in relation to markedly different stances towards the unequal histories and geographies shaped by imperial pasts (see also Schwarz, 2011).

Indeed, the very the posing of the 'multi-cultural question' itself came out of struggles to assert the presence of diverse social relations and shaped contestation over the terms on which places and nations were generated. As Glick Schiller

(2016: n.p.) argues, 'a reflexive conjunctural approach to history and social theory' can 'acknowledge the inextricable links between past nation-state building in Europe and North America, massive violent extractions of wealth from racialized imperial subjects, uprisings in colonial centers and among the colonized, fierce globe-spanning struggles for social justice and the past emergence of democratic reforms and welfare states'. Places, in such accounts, become articulated as key sites through which histories of encounter are staged, articulated and politicized.

This raises important political questions over how to conceptualize the structuring role of such uneven pasts on present social and political relations. In this regard, Bergin and Rupprecht's (2016) arguments about the 'reparative' mode of political and intellectual thought in their insightful introduction to a special issue of *Race and Class* on 'reparative histories' is instructive. They argue that 'reparative histories' can be a way of intervening in 'complex interactions of history, race, agency, memory and trauma' which are 'both essential to and negated by contemporary dominant understandings of racism in the US and Europe'. By 'reparative history', they mean a project which can 'unpack those complex interconnections between past and present in the context of contemporary resistances to racism and the legacies of colonialism' (Bergin & Rupprecht, 2016, p. 5).

Central here is how we think about the political construction of such relations and bring them into contestation. Bergin and Rupprecht (2016, p. 12) argue that the 'concept of the reparative – thought through historically – enables the work of mourning to be connected to the politics of material redress by refusing to understand the history of "race", imperialism and slavery from the vantage point of contemporary progress and reason'. In particular, they argue against 'liberal narratives which would monumentalise and domesticate histories of slavery and colonialism' for the importance of excavating 'histories of resistance, solidarity and collectivity as vital for the now'. In this respect, they contend for the importance of 'acknowledging the presence of black radicalism, black rebellion, anti-colonial struggle' (Bergin & Rupprecht, 2016, p. 12).

As such, their focus on the importance of black resistance and anti-colonial struggle resonates with attempts to shift and politicize the terms on which encounters are experienced and articulated. This opens up different ways of thinking about the diverse ways in which histories and geographies of resistance and struggle have shaped places (see Massey, 1995). Such diverse relations have often been occluded by the normative whiteness that has shaped work on labour and resistance. There are, however, useful if sometimes neglected resources for offering more diverse understandings of the relations between place and solidarity. Thus, Mark Duffield's (1988) fine-grained study of the struggles of Indian foundry workers in the West Midlands of the UK in the post-war period, for example, is keenly attentive to the relations between solidarity, encounters and the political construction of place.

Duffield (1988) traces the emergence of the Indian Shopfloor Movement in foundries in places such as Smethwick which were at the forefront of white racist political movements and discourse in the early to mid-1960s. The contest for the Smethwick constituency in the 1964 general election, for example, was fought by the Conservatives on explicitly racist terrain and was brought to global attention by the visit of Malcolm X to contest their political racism. As Hall et al. (1978, p. 49) argued in *Policing the crisis*, the 'growth of anti-immigrant feeling' through 'the first Commonwealth Immigration Act (1962) restricting immigration, followed by the success of Peter Griffiths on an anti-immigrant ticket at Smethwick in 1964; the reversal of Labour's policy on immigration in 1965' was a 'structuring context' of placed relations and dynamics.

The development of the Indian Shopfloor Movement was supported by the Indian Workers' Association (IWA) which played an important role in promoting and projecting its influence (Duffield, 1988, p. 89). This was allied to the role of the Birmingham IWA which had become 'increasingly active in the campaign against racial oppression' (Duffield, 1988, p. 74). Duffield explores the ways in which Indian foundry workers were both able to secure a marginal place in the labour market in foundries contesting both exploitation by employers, and the exclusionary practices of the main trade unions active in these workplaces, notably the Amalgamated Union of Engineering and Foundry Workers (AUFW). He argues that the 'effect of racial hegemony was to render Indians ineligible for inclusion in the popular forms of resistance to managerial authority' and that 'this exclusion from popular resistance' was not 'corrected by a strong trade union presence in the foundries' (Duffield, 1988, p. 43). Instead he contends that 'trade union practice during the 1950s was aimed precisely at denying blacks the same employment criteria as whites' and colluded with a 'racialized system of managerial authority into which intermediaries, strengthened by the limiting factors of immigration were integrated' (Duffield, 1988, p. 43).

The ways in which these struggles over place and racial hegemony were central to different organizing practices emerges through Indian shopfloor-led strikes such as one at the Coneygre Foundry, Tipton, in 1967 which was occasioned by the decision by management to make 21 Indians redundant and to ignore the offer of the 'Indian workforce to accept work sharing and hence reduced wages in place of these redundancies' (Duffield, 1988, p. 87). As Duffield (1988, p. 87) notes, during the dispute 'management was able to use the 150 white workers, none of whom joined the strike, in an attempt to keep the foundry going' and notes that many of the workers who crossed picket lines were AUFW members. 'Outside of the company, the main support for the strike came from other Asian workers' together with 'students from Birmingham and Aston Universities who showed their solidarity by standing on the picket line'. The IWA helped to 'supply transport, placards and other logistical support' and 'issued a press statement urging all workers in the company to join the Asians' and make it 'a strike of international solidarity'.

The picket line is positioned here as a key site of encounter where articulations of solidarity can reconfigure political constructions of place. Thus, Diarmaid Kelliher (2017) has demonstrated how the picket lines during the Grunwick strike 'in the London Borough of Brent' – this 'strike at a photo processing plant, largely led by Asian women including Jayaben Desai' – became an important site of solidarity and multi-ethnic encounters. Thus, he notes that in mid-1977 '"miners, dockers, engineers and building workers swelled the picket" to thousands strong' (Kelliher, 2017, p. 10). These relations and connections shaped diverse solidarities in ways which pre-figured the important involvement of groups such as Lesbians and Gays Support the Miners during the 1984–1985 Miners' Strike which have been celebrated in the film *Pride*.

This emphasizes the importance of viewing solidarities as produced through relations and as shaping encounters, rather than merely pre-existing them. It also stresses the dynamic relations between solidarities, place and connections that stretch beyond place (see Massey, 2007). These engagements with political struggles over place, 'race' and labour in post-war Britain emphasize that unpacking the construction of place can be necessary to engage with the terms on which encounters are generated, articulated and politicized. They also assert the need to think about how we integrate such histories and geographies of resistance more directly into the way we approach and contextualize encounters. This is significant as such histories and geographies of contestation have often been held at some distance from work on actually existing negotiations of encounters.

Thus, Gill Valentine's influential critique of work on geographies of encounters asserts the importance of recognising that encounters 'never take place in a space free from history, material conditions and power' (Valentine, 2008, p. 333). She argues that 'by celebrating the potential of everyday encounters to produce social transformations', 'contemporary discourses about cosmopolitanism and new urban citizenship' 'potentially allow the knotty issues of inequalities to slip out of the debate' (Valentine, 2008, p. 333). While her caution against 'romanticising encounters' is important, there remain tensions in the ways in which she envisions encounters. Thus, her work gives a rather static, disembodied sense of encounters and her analysis tends to position whiteness as given. Thus, she argues that the 'reason that such individual everyday encounters do not necessarily change people's general prejudices is because they do not destabilize white majority community based narratives of economic and/or cultural victimhood' (Valentine, 2008, p. 333).

To challenge the terms on which white majority narratives are constructed, however, necessitates thinking about the ways in which whiteness is actively produced and performed through racialized power relations. As Akwugo Emejulu (2016: n.p.) has noted in relation to debates around 'race' and Brexit, 'whiteness' becomes 'recast as witness to racism, but without any imperative to dismantle white supremacy, the system of racial hierarchy remains firmly in place, with whiteness preserved, unchallenged and intact'. By abstracting such encounters from the unequal terms on which whiteness is shaped then, Valentine occludes such encounters from the relations of power, shaped through productions of whiteness. The role of particular constructions of whiteness is important in shaping terms on which place is envisioned and generated. In particular, it is necessary to think about how whiteness shapes and produces some encounters on decidedly unequal terms.

In this regard, a more lively sense of the formation of 'encounters' is central to Doreen Massey's (2005) account of space as a site of the co-existence of different trajectories. In her own words, her account of place 'as an ever-shifting constellation of trajectories poses the question of our throwntogetherness'

(Massey, 2005, p. 151). Massey's account of 'throwntogetherness' does not imply an easy pluralism or negotiation, rather she envisions such processes as resulting as much in political antagonism and contestation as productive engagement between unlike actors. In this sense, she does not, however, fetishize diversity, rather insisting on the need to understand the unequal relations that shape relations between diverse groups in the city. She helpfully moves beyond the limits of accounts which see diversity as a 'simple plurality'. Arguing of London, for example, she positions it is a constellation of 'trajectories with different resources, distinct dynamics (and strengths in the market) and temporalities, which have their own directions in space-time, and which are quite differently embedded within "globalisation"' (Massey, 2005, p. 157).

What is important here is the way that her work is alive to the many different ways in which the meeting of such trajectories can play out and be negotiated. As Laura-Jane Nolan (2015) has usefully demonstrated in her PhD research on Kinning Park Complex, a community centre on the Southside of Glasgow, while Massey's account of throwntogetherness is rather abstract, it can be used to understand the different ways in which encounters between unlike actors can be negotiated. Through detailed ethnographic engagements, Nolan's work elaborates the different kinds of outcomes of diverse trajectories, while also drawing attention to the tensioned processes through which such practices developed/negotiated. Nolan also locates such encounters in relation to conjunctural processes of austerity and neoliberalization and the way they impact on – and are negotiated through – placed relations and activities. In this regard, her arguments resonate with the particular project that is central to this book.

Diverse modalities of solidarity and encounter

In their introduction to this collection, Oosterlynck et al. note that the 'solidarities we are looking for in this volume do not presuppose integration into a predefined community, nor the necessity of historical time to build up social capital between diverse citizens, but require a willingness to negotiate the diversity of people, practices, claims and networks that affect a particular place'. One of the important contributions of the collection is the way that the chapters give a sense of what might be thought of as 'diverse modalities of solidarity and encounter'. The chapters give a vibrant sense of the practised, embodied articulations of encounters, their relations to the construction of practices of solidarity and how they can reconfigure place-based relations.

As a result, one of the strengths of the collection is the attention the chapters pay to the different sites, spaces and relations through which different solidarities are shaped and articulated. The book is enriched by a whole range of engagements with encounters and place-making practices. The book usefully highlights aspects of place-making which are often ignored or hidden. Thus, Michela Semprebon and Martina Valsesia explore forms of self-building projects by migrants as a response to 'diminishing public investment in social housing, enduring forms of discrimination towards migrants and the multiplication of forms of evictions' in contemporary Italy. They argue that engagement with 'the actual process of self-building has nurtured solidarities, how these solidarities have supported the realization of the projects and how solidaristic bonds have been challenged throughout the projects'.

As well as drawing attention to often marginalized interventions in the politics of place, the contributions draw attention to neglected techniques through which solidarities are constructed. Thus, Mervyn Horgan's discussion of strangers and solidarity in urban life insightfully unsettles assumptions that likeness and proximity are the only routes towards the construction of solidarities. He argues that indifference, rather than being incompatible with the articulation of solidarity, can at times be important in its production. His call for attending to the 'uneven and asymmetrical distribution and enactment of indifference' is crucial because it suggests the importance of recognizing the diverse modalities of interaction that shape the formation of solidarities.

The collection's sensitivity to placed dynamics also helps to enliven debates through engaging with a whole different repertoire of practices through which diverse solidarities are enacted. The chapters give a sense of the different ways in which intentional attempts to produce and shift the terms of encounters can be negotiated. In their fascinating chapter on challenges to entrenched ethnic divisions between 'the Baka and the Bantu in East Cameroon', Awuh and Spijkers examine the use of collaborative football games and latrine digging to challenge 'stereotyping and lack of positive interaction between Baka and Bantu'. They note that the football programme and the communal latrine digging exercise laid the foundation for 'prolonged everyday place-based practices in which the Baka jointly engaged with their Bantu neighbours in social activities'. They underline that 'such place-based solidarities no longer presuppose a set of shared norms and practices, but a willingness to engage with the diversity of people living in the periphery of the Dja Reserve area'.

In Walter et al.'s chapter, we see how particular ways of envisioning solidarities through faith-based practices shaped by trajectories linked to the civil rights movement are articulated through particular spatial practices; in this instance a strategy of 'intentional neighbouring' with those in poor and deprived neighbourhoods. They contend that 'the spatial solidarity of intentional neighbouring shapes and reinforces other sources of solidarity such as interdependence, shared values and common struggles that could lead to collective action and new forms of living together in difference'.

In similar terms, Schuermans and Debruyne explore the strategies used by a school in Leuven, Belgium, to negotiate 'superdiversity' in positive ways. They caution, however, that while 'progressive forces within the school count on intercultural encounters outside the school territory as a means to nurture solidarity among more conservative colleagues, such encounters alone are not sufficient to reach these positive effects. Instead, it needs to be acknowledged that many of the encounters and solidarities at Mater Dei are strongly mediated by personal histories, pre-existing values and struggles for social justice.'

There is, in this regard, an important recognition in the collection that intentional attempts to shape solidarities and diverse communities can also be fraught with their own power relations and tensions. Mandy de Wilde's engagement with the dynamics of the Neighbourhood Circle, in a 'superdiverse' Amsterdam neighbourhood, for example, interrogates the terms on which 'residents enact solidarities' through 'multi-cultural spectacle'. She emphasizes that rather than producing a uniformly generous notion of the neighbourhood, Dutch volunteers actively contest the 'use of urban public space by both immigrant women and the creative class'. As she notes, 'Dirk and his fellow volunteers perceive the diversity and the accompanying public recognition of multicultural encounters as problematic. To them it affects their ability to publicly articulate their Dutch identity through practices such as organizing encounters and also to receive public recognition for it.'

What is important here is the recognition of how solidarities can entrench, challenge and rework relations of power within particular places (Featherstone, 2012). This raises the importance of thinking through the terms on which such solidarities are generated. One way of engaging with diverse modalities of solidarity is to unpack ways of thinking about the terms on which relations and connections are shaped. Thus, in her chapter Helen Wilson argues for the importance of thinking about the dynamics between friendships, intimacy and solidarity. She usefully contends that 'attending to the *spaces* in which solidarities and friendships are built, [. . .] challenges the assumptions of spatial boundedness that have shaped understandings of solidarity and its supposed links to communities of similitude'. Through doing so, her account seeks 'to move away from grand and linear narratives of struggle, to offer an account that foregrounds the multiple scales, gestures and often "quiet politics" (Askins, 2015) involved'. As such, her chapter 'highlights the often-overlooked and less spectacular ways in which social action gets done'. As Wilson (forthcoming) argues elsewhere, in a discussion of Alex Vasudevan's (2015) work on squatting in Berlin, such a focus on 'encounters as part of a range of performative techniques' can destabilize 'the border between politics and the everyday, artist and audience, and activist and non-activist'.

In similar terms, Bruno Meeus's discussion of Romanian migrants in Brussels is keenly alive to the ambiguities of solidarity, noting some of the different uses that solidarity can be put. Thus, he notes of a Romanian Catholic priest in Brussels that while mobilizing forms of ethnic solidarity, at the same time he seeks to purify the Romanian presence in Brussels by mobilizing an ethnicized morality of hard work, defining such 'hardworking' Romanians against the Roma. Through examining conditions of cleaning work at a Brussels railway station, Meeus locates Romanian workers in relation to processes of precarity, exploitative labour relations and migration. What is particularly prescient and insightful here is the way in which Meeus traces the way in which 'the situational we-ness that developed in the workplace had been constructed beyond lines of ethnic division' noting the way a petition 'written by one of the railway employees, framed the struggle as a shared struggle against new attacks on the railway employee status'.

This breaks down the way in which storying of migrant labour geographies can at times reproduce 'migrant'/'native' worker divisions in rather stark terms. Wills et al. (2010, p. 2), for example, by counterposing 'migrants and established working class communities' play in to an imaginary that sees working-class communities as fully formed before migration rather than as the ongoing products of diverse trajectories. In this regard, it is crucial to note how spaces of organizing and encounter have shaped and been shaped by solidarities which articulate new ways of relating and struggling against unequal ways of generating place. Such articulations are particularly important in a context where labour organizing and the activity of some social movements have been articulated in ways that intensify rather than challenge or circumvent such divisions. In the final section, I will suggest why the formation of such strategies is such a crucial political task in terms of the current post-crisis conjuncture.

Solidarities and encounters in a conjuncture of crisis and austerity

In his stark essay 'Fascism then and now' in the *2016 Socialist Register*, Geoff Eley argues that dynamics of disaffection around contemporary economic change have increasingly been racialized. He contends that '[w]hat gives the new politics of the right traction' is that such 'change produces exactly the multifarious anxieties about boundaries whose interconnectedness xenophobia then readily cements' (Eley, 2016, p. 111). As Stuart Hall (1978, p. 31) argued in the late 1970s, 'race' can function as a key 'lens through which people come to perceive that a crisis is developing' and can be 'the framework through which the crisis is experienced'.

Such divisive, and frequently racialized constructions of crisis have been central to the dominant ways in which the post-2008 crisis has been articulated and politicized. A key challenge of conjunctural analysis is to understand the way migration has become a central referent point in relation to such debates and the ways in which this has been worked through particular constructions of place and nation. This raises the question of what the implications of this are for understanding the futures of solidarities and encounters and terms of contemporary conjuncture(s). In this respect, I wish to draw attention to three particular processes and the divisive geographies through which they are constituted which 'exert pressure and set limits' on the formation of diverse solidarities and generous articulations of place. In this regard we see how solidarities are constituted within, but are not determined by, particularly structural constraints and forces.

First, it is important to recognize that logics of austerity have worked through divisive geographies both within places and through the mobilizing of spatial divisions such as between Northern and Southern Europe. A concerning aspect of these divisive geographies has been the ways in which various forms of left-wing and trade union organizing has been structured around such divisions rather than challenging them, thus high profile disputes have mobilized around slogans like 'British Jobs for British Workers' (Ince et al., 2015).

In a recent discussion of trade union organizing in the Danish construction sector, Oscar Garcia Agustín and Martin Jørgensen (2016, p. 156) have argued that a 'critique of the neoliberal model and its dominant role in Europe was displaced by critique of the EU principle of free movement, which was perceived as a risk for the Danish welfare state (or the so-called "flexicurity" model) and Danish workers' rights and decent wages'. This mobilization 'gave the far-right Danish People's Party (Dansk Folkeparti) an opportunity to target the debate about the threat represented by Eastern European workers (Johansen, 2014)' (Agustín & Jørgensen, 2016, p. 156). In line with Meeus's chapter in this volume, however, they point at 'different cultures of union organising in relation to migrant workers in Norwegian and Danish construction sectors'. They observe for example that Norwegian trade unionists have refused 'to take action against work places with foreign workers because this creates a division between "us" and "them"'.

Second, this emphasizes that is not just the far-right who have sought to exploit and intensify such divisions. The response of trade unions and centre-left parties has often been problematic. In part, this has stemmed from a failure to adequately challenge hostile discourses around immigration and migrant labour. But, perhaps most seriously, it has been reticent to articulate positive narratives about the construction of places as sites of multiculture. In this regard, it is necessary to challenge the kind of bleached, whitened accounts of the past which are often reproduced by the left as well as the right. This speaks to why reinvigorating debates around progressive articulations of place in relation to diverse articulations of solidarity both within and beyond place is a crucial terrain of struggle for contemporary left politics.

Third, it is essential to engage with the uneven impacts of austerity and place. Processes of austerity in the UK have had key racialized, gendered and classed impacts, with a disproportionate impact on BME women and their organizations (Vacchelli et al., 2015). It is essential here to articulate a challenge to – and to attempt to transcend – the racialized, gendered and classed divisions that neoliberal strategies and precarious working practices thrive on and intensify. As Kate Derickson's (2016, pp. 1–2) account of the 'synthesis' of 'racialized dispossession, austerity urbanism, and broken windows policing' testifies, such divisions are being generated in different contexts.

Recent work has demonstrated how Greece has emerged as one of the countries at the sharpest end of regressive policies of austerity. Athina Arampatzi and Lazaros Karaliotas have both demonstrated in different ways that social and political movements on the Greek left have articulated important challenges to such divisive imaginaries, and have shown that possibilities for generating solidarities have potential to shift the terms of debate on the terms which places are generated and articulated. Thus, Arampatzi argues that such alternative practices can 'constitute an empowering process of solidarity-making "from below"', and open up spaces for the practice of bottom-up democratic politics vis-a-vis austerity, a 'politics of fear and crisis' (Arampatzi, 2016, p. 1). By challenging the givenness of place and solidarity, Karaliotas's work is particularly useful in the way that it highlights how the production of solidarities can refigure the existing political terrain and relations (Karaliatos, 2016).

In this vein, these challenges all emphasize some of the tensions of the contemporary conjuncture for diverse articulations of solidarity in place. They also, however, underline why the production, maintenance and articulation of solidarities matter. This book has usefully emphasized some of the different political skills needed to shape such diverse placed solidarities. Further, it has suggested some of the diverse textured and lived processes through which such solidarities are achieved. These are crucial and urgent tasks. For as Agustín and Jørgensen (2016, p. 164) argue: 'What is at stake here is the constitution of a social Europe in times when neoliberal policies are undermining social policies inherited from welfare states. Migrants are consequently necessary allies in the attempt to build the social dimension that Europe is lacking. Otherwise, inequality and exclusion will increase.'

Notes

1 The song is referred to by its nickname 'Auld Reekie'.
2 References to the lyrics of the song are in line with the version sung by Dick Gaughan, the foremost interpreter of the song, on his fine album *Handful of Earth*, Topic Records, 12TS419, 1981.
3 The song was included in Ord's collection of *Bothy songs and ballads* (see Ord, 1995, p. 387), ironically given that John Ord was a Superintendent of the Glasgow Police Force!

References

Agustín, O. G., & Jørgensen, M. B. (2016). For the sake of workers but not immigrant workers? Social dumping and free movement. In O. G. Agustín & M. B. Jørgensen (Eds.), *Solidarity without borders: Gramscian perspectives and civil society alliances* (pp. 150–168). London: Pluto Press.

Arampatzi, A. (2016). The spatiality of counter-austerity politics in Athens, Greece: Emergent 'urban solidarity spaces'. *Urban Studies*, published online before print, doi: 10.1177/0042098016629311.

Askins, K. (2015). Being together: Everyday geographies and the quiet politics of belonging. *ACME: An International Journal for Critical Geographies*, *14*(2), 470–478.

Bergin, C., & Rupprecht, A. (2016). History, agency and the representation of 'race': An introduction. *Race and Class*, *57*(3), 3–17.

Derickson, K. D. (2016). Urban geography II: Urban geography in the Age of Ferguson. *Progress in Human Geography*, published online before print, doi:10.1177/030913 2515624315.

Duffield, M. (1988). *Black radicalism and the politics of de-industrialisation: The hidden history of Indian foundry workers.* Aldershot: Avebury.

Eley, G. (2016). Fascism then and now. *Socialist Register*, *52*, 91–117.

Emejulu, A. (2016). *On the hideous whiteness of Brexit: 'Let us be honest about our past and our present if we truly seek to dismantle white supremacy'*. Retrieved from www. versobooks.com/blogs/2733-on-the-hideous-whiteness-of-brexit-let-us-be-honest-about-our-past-and-our-present-if-we-truly-seek-to-dismantle-white-supremacy (last accessed 11 August 2016).

Featherstone, D. (2012). *Solidarity: Hidden histories and geographies of internationalism.* London: Zed Books.

Glick Schiller, N. (2016). The question of solidarity and society: Comment on Will Kymlicka's article: 'Solidarity in Diverse Societies'. *Comparative Migration Studies*, *4*(6), doi:10.1186/s40878-016-0027-x.

Hall, S. (1978). Racism and reaction: A public talk arranged by the British Sociological Association and given in London on 2 May 1978. In Commission for Racial Equality (Eds.), *Five views of multi-racial Britain* (pp. 23–35). London: Commission for Racial Equality.

Hall, S. (1988). *The hard road to renewal: Thatcherism and the crisis of the left.* London: Verso.

Hall, S. (2000). Conclusion: The multi-cultural question. In B. Hesse (Ed.), *Un/settled multiculturalisms: Diasporas, entanglements, transruptions* (pp. 209–241). London: Zed Books.

Hall, S., Critcher, C., Jefferson, T., Clarke, J., & Roberts, B. (1978). *Policing the crisis: 'Mugging', the state and law and order.* London: Macmillan.

Ince, A., Featherstone, D., Cumbers, A., MacKinnon, D., & Strauss, K. (2015). British jobs for British workers? Negotiating work, nation, and globalisation through the Lindsey Oil Refinery disputes. *Antipode, 47*(1), 139–157.

Karaliatos, L. (2016). Equals in solidarity: Thessaloniki's Social Centre – Immigrants' Place. Unpublished paper, available from author.

Kelliher, D. (2017). Constructing a culture of solidarity: London and the British coalfields in the long 1970s. *Antipode*, published online before print, DOI: 10.1111/anti.12245.

Massey, D. (1995). Places and their pasts. *History Workshop Journal, 39*(1), 182–192.

Massey, D. (2005). *For space.* London: SAGE.

Massey, D. (2007). *World city.* Cambridge: Polity Press.

Nolan, L.-J. (2015). *Space, politics and community: The case of Kinning Park Complex.* Unpublished PhD thesis, University of Glasgow.

Ord, J. (1995). *Ord's bothy songs and ballads of Aberdeen Banff and Moray Angus and the Mearns.* Edinburgh: John Donald.

Vacchelli, E., Kathrecha, P., & Gyte, N. (2015). *Is it really just the cuts? Neo-liberal tales from the women's voluntary and community sector in London. Feminist Review, 109*, 180–189.

Valentine, G. (2008). Living with difference: Reflections on geographies of encounter. *Progress in Human Geography, 32*(3), 323–337.

Vasudevan, A. (2015). *Metropolitan preoccupations: The spatial politics of squatting in Berlin.* Oxford: Wiley-Blackwell.

Wills, J., Datta, K., Evans, Y., Herbert, J., May, J., & McIlwane, C. (2010). *Global cities at work: New migrant divisions of labour.* London and New York: Pluto Press.

Wilson, H. (forthcoming). Occupation, intimacy and the possibilities of encounter. *Urban Studies.*

Index

Taylor & Francis eBooks

Helping you to choose the right eBooks for your Library

Add Routledge titles to your library's digital collection today. Taylor and Francis ebooks contains over 50,000 titles in the Humanities, Social Sciences, Behavioural Sciences, Built Environment and Law.

Choose from a range of subject packages or create your own!

Benefits for you
- ›› Free MARC records
- ›› COUNTER-compliant usage statistics
- ›› Flexible purchase and pricing options
- ›› All titles DRM-free.

| REQUEST YOUR **FREE** INSTITUTIONAL TRIAL TODAY | **Free Trials Available** We offer free trials to qualifying academic, corporate and government customers. |

Benefits for your user
- ›› Off-site, anytime access via Athens or referring URL
- ›› Print or copy pages or chapters
- ›› Full content search
- ›› Bookmark, highlight and annotate text
- ›› Access to thousands of pages of quality research at the click of a button.

eCollections – Choose from over 30 subject eCollections, including:

Archaeology	Language Learning
Architecture	Law
Asian Studies	Literature
Business & Management	Media & Communication
Classical Studies	Middle East Studies
Construction	Music
Creative & Media Arts	Philosophy
Criminology & Criminal Justice	Planning
Economics	Politics
Education	Psychology & Mental Health
Energy	Religion
Engineering	Security
English Language & Linguistics	Social Work
Environment & Sustainability	Sociology
Geography	Sport
Health Studies	Theatre & Performance
History	Tourism, Hospitality & Events

For more information, pricing enquiries or to order a free trial, please contact your local sales team:
www.tandfebooks.com/page/sales

 Routledge Taylor & Francis Group | The home of Routledge books

www.tandfebooks.com